DESIGN OF HIGH FREQUENCY FILTERS

RFI SUPPRESSION AND DIELECTRIC MATERIALS

Edited by T. Kawai

Written by faculty members of the
Georgia Institute of Technology

W. B. Warren, Jr.

J. R. Walsh, Jr.

H. W. Denny

C. S. Wilson

J. A. Hart

Wexford Press

ISBN 978-1-934939-62-8

TABLE OF CONTENTS

Contents	Page

TABLE OF CONTENTS (Concluded)

Contents   Page

LIST OF FIGURES

LIST OF FIGURES (Continued)

LIST OF FIGURES (Concluded)

LIST OF TABLES

1. INTRODUCTION

Low-pass and bandpass filters offer a straightforward means of improving the interference characteristics of transmitters and receivers. The use of such filters is an attractive method of protecting equipment, since the additional interference rejection they provide can generally be obtained without the necessity of making extensive modifications to the equipment being protected.

In many instances, conventional filters have serious spurious responses. In addition, many of these filters have excessively large physical dimensions, particularly at high frequencies.

This book is primarily concerned with techniques for designing and producing filters with low spurious response levels and with small physical dimensions. The techniques developed involve the use of dielectric materials with frequency sensitive loss characteristics and slow propagation velocities.

2. DISCUSSION

2.1 Control of Filter Spurious Responses

If interference suppression filters are to be completely effective in supplying the required interference rejection, it is necessary that the attenuation of the filters be high at frequencies outside the desired passband. Unfortunately, many conventional reactive filters commonly used to provide interference rejection do not meet the high stop-band attenuation requirement. Instead, they exhibit regions of relatively low attenuation which sometimes coincide with a spurious response or spurious emission of the equipment being protected. In many instances, the inherent reactance versus frequency behavior of the elements from which conventional filters are constructed strongly favors the production of multiple passbands in the filters' transfer functions. For example, the inductances and capacitances of conventional short line coaxial filters are formed by short transmission line sections of the proper characteristic impedance. Since the reactance of these transmission lines are periodic functions of frequency, there is an inherent periodicity in the passband and stop-band characteristics of such filters. The periodic frequency behavior of a typical low-pass filter attenuation function is shown in Figure 1. For many applications, the rapid cutoff slope and the relatively wide stop-band between the first cutoff frequency and the first spurious passband of such a filter are quite satisfactory, but for RFI applications the multiple spurious passbands of such a filter are unacceptable.

2

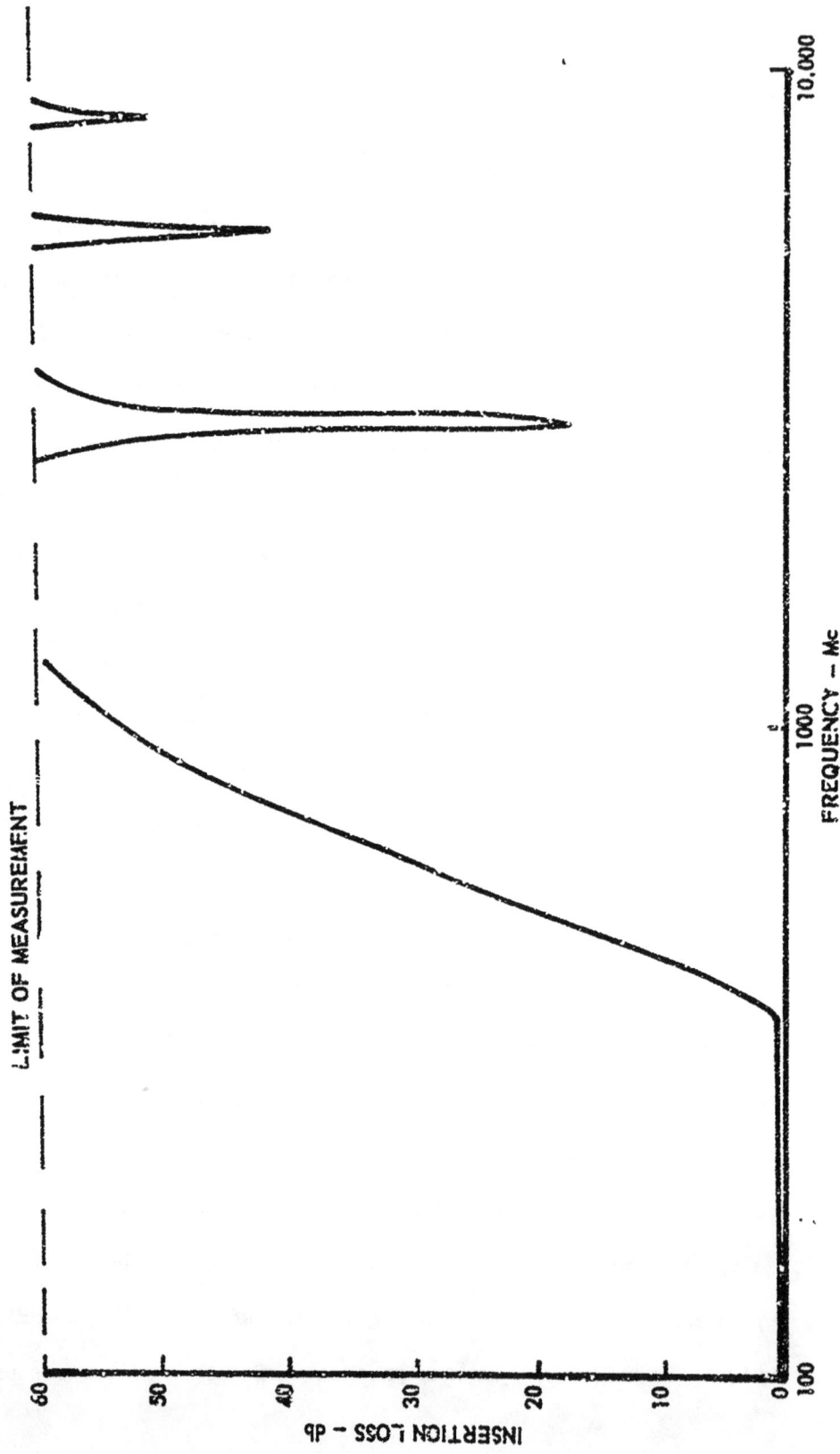

Figure 1. Typical Low-Pass Attenuation Function.

One method of avoiding these spurious passbands is to employ a dissipative mechanism to produce the desired low-pass characteristic. The required frequency sensitive attenuation can be provided by a coaxial line in which the dielectric space between the center and outer conductors has been filled with a material whose attenuation constant is an increasing function of frequency.

The frequency sensitivity of the attenuation constant of the dielectric material arises primarily from the inability of the elementary electric and magnetic dipoles which make up the dielectric material to follow instantaneously the electric and magnetic fields applied to the dielectric. The effect of the time lag of the electric and magnetic dipoles with respect to the applied fields can be expressed in terms of a complex permeability and complex permittivity of the dielectric. Hence,

$$\mu^* = \mu_o \left(\mu_r' - j\mu_r'' \right) \quad ; \tag{1}$$

and

$$\epsilon^* = \epsilon_o \left(\epsilon_r' - j\epsilon_r'' \right) \quad . \tag{2}$$

To illustrate how the complex permittivity can account for a loss component, consider the impedance of a simple parallel plate capacitor The impedance, Z_c, is

$$Z_c = \frac{1}{j\omega C} = \frac{1}{j\omega A \epsilon^*/d} = \frac{d}{j\omega A \epsilon_o (\epsilon_r' - j\epsilon_r'')} \quad , \tag{3}$$

4

where A is the area of one of the plates and d is the spacing between the plates. Rationalizing the denominator gives

$$Z_c = \frac{(\epsilon_r'') \, d}{\omega A \epsilon_o \left[(\epsilon_r'')^2 + (\epsilon_r')^2 \right]} - j \frac{(\epsilon_r') \, d}{\omega A \epsilon_o \left[(\epsilon_r'')^2 + (\epsilon_r')^2 \right]} \quad . \tag{4}$$

This impedance has a resistive component of

$$R_e \left\{ Z_c \right\} = \frac{(\epsilon_r'') \, d}{\omega A \epsilon_o \left[(\epsilon_r'')^2 + (\epsilon_r')^2 \right]} \quad , \tag{5}$$

which can account for any dissipation in the dielectric of the capacitor. Likewise, for an inductance, the impedance is proportional to the permeability of the magnetic medium on which the coil is wound. Consequently,

$$Z_L = j\omega L = j\omega k \mu^* = j\omega k \, (\mu_r' - j\mu_r'') = j\omega k \mu_r' + \omega k \mu_r'' \quad , \tag{6}$$

where k is a proportionality constant. Again, there is a resistive term, $\omega k \mu_r''$, which accounts for dissipation in the magnetic medium.

The connection between μ^* and ϵ^* and the magnetic and electric dipoles of a dielectric medium is treated in more detail in Appendix A, where the propagation factor of a TEM wave in a medium with complex permeability and complex permittivity is derived and the attenuation constant, α, of the TEM wave is shown to be

5

$$\alpha = \left\{ (12.85)(10^{-10}) \ f \sqrt{\mu_r' \ \epsilon_r'} \right\}$$

$$\left\{ \sqrt{(1 + \delta_\mu^2)(1 + \delta_\epsilon^2)} - 1 + \delta_\mu \delta_\epsilon \right\}^{1/2} \ db/cm \quad , \tag{7}$$

where f is the frequency in cps, and δ_μ and δ_ϵ are the magnetic and dielectric loss tangents respectively. The quantities δ_μ and δ_ϵ are defined by

$$\delta_\mu = \tan^{-1} \left\{ \frac{\mu_r''}{\mu_r'} \right\} \quad , \tag{8}$$

and

$$\delta_\epsilon = \tan^{-1} \left\{ \frac{\epsilon_r''}{\epsilon_r'} \right\} \quad . \tag{9}$$

The factor f, in Equation (7), indicates a rising attenuation with increasing frequency as long as the parameters of the dielectric material remain constant. However, the relative dielectric constant and relative permeability of most materials are decreasing functions of frequency, and an overriding increase in the loss tangents must occur if Equation (7) is to reproduce the experimentally observed low-pass attenuation function. Table 1 gives some typical values[1] of the constants of a dielectric material formed by suspending an iron powder in an epoxy binder. If the data of Table 1 is substituted in Equation (7), the resulting loss versus frequency function is that shown in Figure 2. Also shown in Figure 2 is the measured loss of a coaxial line section filled with the iron dielectric

6

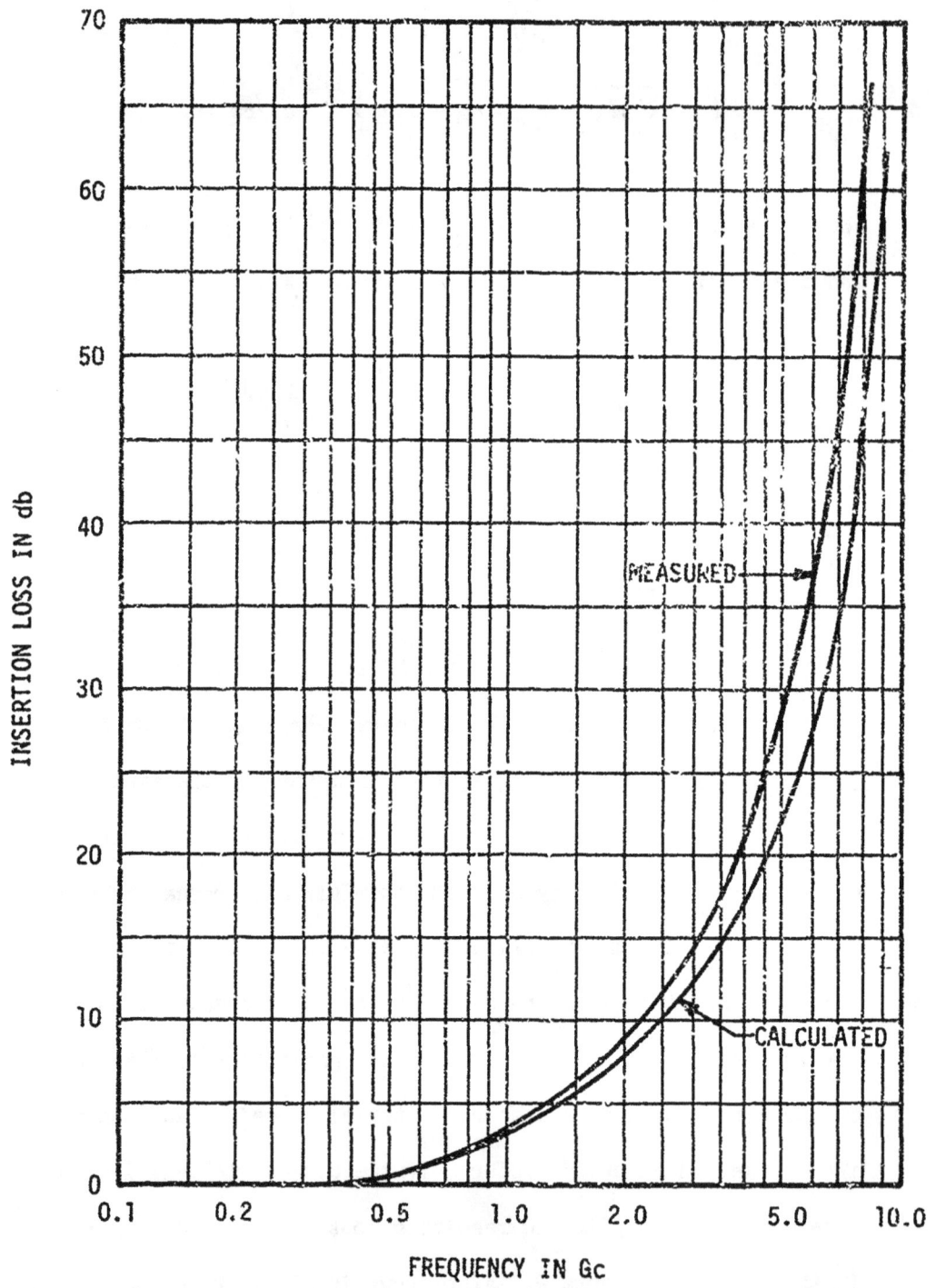

Figure 2. Calculated and Measured Attenuation.

material. The experimental and calculated values are in good
agreement at low values of attenuation, with the divergence between
measured and calculated values increasing as the attenuation rises.

The divergence of the two curves of Figure 2 can be assigned,
in part, to errors in the values of the material constants given
in Table 1. In addition, some of the excess of the measured attenuation over that calculated can be attributed to the increasingly
poor impedance match at the input to the lossy line section as frequency increases. Since the characteristic impedance of the
coaxial line is proportional to $(\mu^*/\epsilon^*)^{1/2}$, it can be inferred from
the data of Table 1 that a significant variation of both the magni-
tude and angle of the characteristic impedance takes place over the
frequency range 0.1 Gc to 10 Gc, with a resulting increase in the
measured insertion loss of the coaxial line due to reflections at
the input and load terminals of the line.

TABLE 1

MATERIAL CONSTANTS

Frequency	δ_μ	δ_ϵ	μ	ϵ
100 mc	0	0.0375	4.7	20.0
1000	0	0.043	4.0	19.0
1500	0.16	0.044	3.6	19.0
3000	0.2	0.045	3.4	19.0
10000	0.5	0.048	2.2	18.5

8

2.1.1 Loss Characteristics of Dielectric Materials

A large amount of data was collected on a variety of materials to determine their suitability for use in lossy transmission line filters. A standard test procedure was used to facilitate the intercomparison of the loss properties of the materials tested. The standard test was conducted by measuring the attenuation versus frequency characteristics of a standard length of coaxial transmission line whose dielectric space had been filled with the lossy dielectric material. This method of testing was selected because it closely approximates the configuration in which the lossy dielectric material would be used in many practical applications. The standard coaxial test section used consisted of a 7.2 cm long brass outer conductor with an inside diameter of 0.56 inches and a solid copper center conductor 0.102 inches in diameter. Type N coaxial fittings were placed at each end of the test section to provide connections to a 50 ohm insertion loss measuring system. Procedures for the preparation of the lossy dielectric material as well as for construction of the transmission line sections are treated in more detail in Section No. 2.1.10.

The loss data obtained on the various materials tested is presented in the form of attenuation versus frequency curves. These curves permit the loss per unit length of transmission line at a given frequency to be easily calculated by dividing the loss read from the curves by the length of the test sample. Since the μ_r^*

and ϵ_r^* are different for the several materials tested, the
characteristic impedance of the standard test sections varies, and
a variable amount of reflection loss is introduced at the interface between the 50 ohm test system and the test specimen. At
frequencies where the dissipation losses in the transmission line
are large, the effect of these reflection losses is relatively
minor and the attenuation constant of the material can be obtained
directly from the curves. However, at frequencies where the dissipation losses are small, the reflection losses may predominate and
the true propagation constant of the dielectric material cannot be
determined directly from the standard attenuation curves. In this
case, the correct value of the attenuation constant can be obtained
by subtracting the reflection loss from the value shown on the
attenuation curve for the material in question. The proper amount
to be subtracted can be obtained from the relation

$$L_R = 40 \log_{10}\left(\sqrt{\mu_r^*/\epsilon_r^*} + 1/2\right) - 10 \log_{10}\left(4\mu_r^*/\epsilon_r^*\right) \text{ db} \quad . \quad (10)$$

Equation (10) accounts for the reflection loss at both the input
and output interfaces of the test sample but does not take into
account the effect of multiple reflections. For the values of the
attenuation constant encountered in the dielectric materials studied,
the effect of multiple reflections is negligible. A derivation of
Equation (10) is given in Appendix B.

2.1.1.1 Carbonyl Iron

The curves shown in Figure 3 summarize the results
obtained for a dielectric material composed of carbonyl iron and
epoxy. The ratios given on each curve are the ratios by weight of
iron to epoxy. By varying the iron to epoxy ratio, the attenuation
at a given frequency can be varied over a wide range. The carbonyl
iron material is well suited for use in lossy filters whose passbands extend to the UHF region since the loss of the lossy transmission can be made low over the passband region by using a
sufficiently short line section or by limiting the mix ratio to a
low value. Although the loss increases in rough proportion to the
amount of iron used, the attenuation is not, in general, a linear
function of the mix ratio, but varies in the manner shown in
Figure 4. From a practical viewpoint, mix ratios in the range
from 2:1 to 6:1 are the most useful because the iron and epoxy
mixture becomes quite thick at high mix ratios, while the epoxy
losses tend to overshadow the iron losses at ratios much below 2:1.

The carbonyl iron powder used in forming the test sections
was Grade "E" and was obtained from the General Aniline and Film
Corporation. The average particle size of the "E" grade carbonyl
iron powder is in the neighborhood of 8 microns. Two other grades
of carbonyl iron containing slightly smaller particles are available. These two grades are designated as Grade SF and Grade TH.
For both these grades, the average particle size is about 4
microns. The use of smaller particles in the lossy dielectric

11

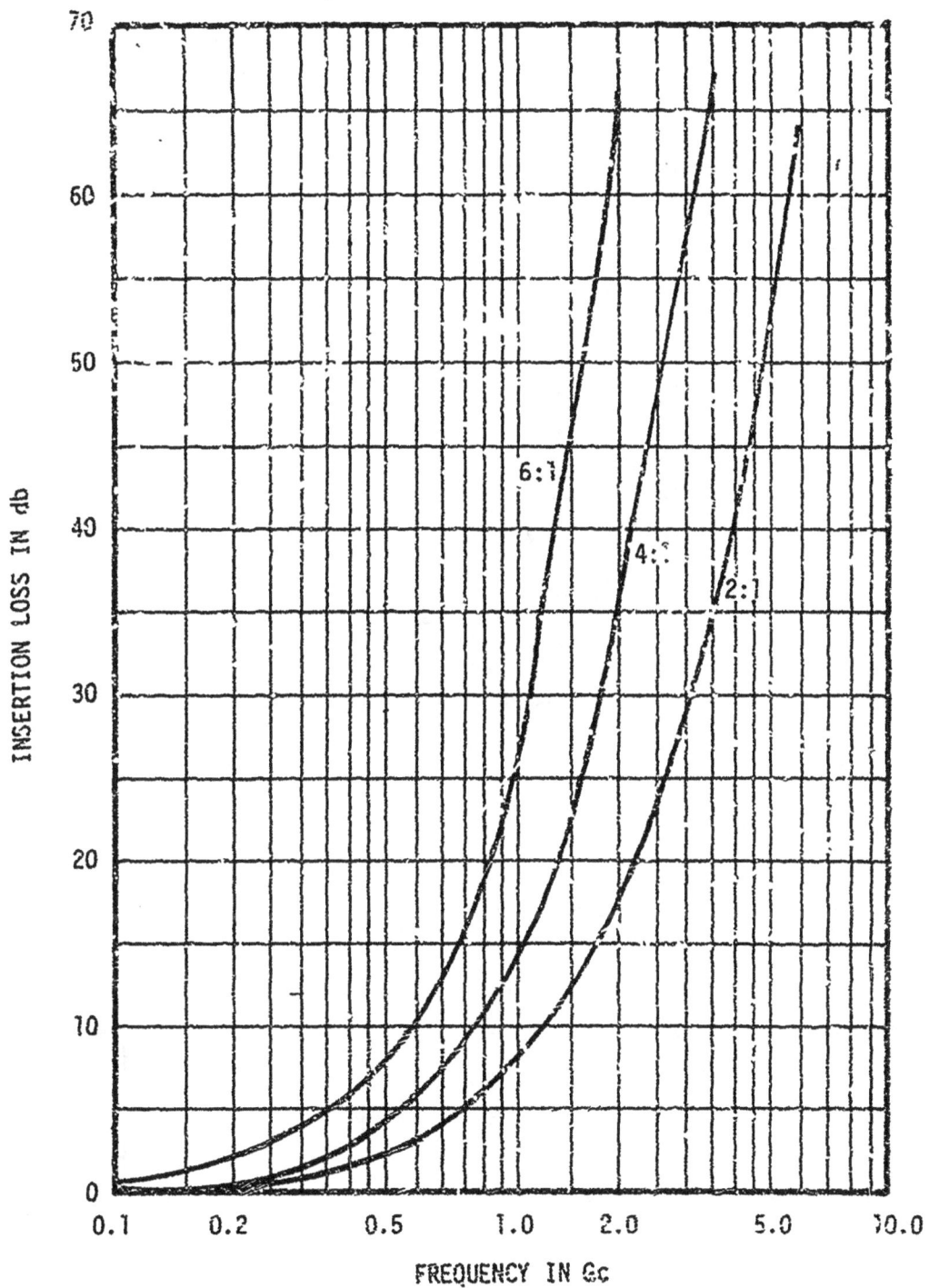

Figure 3. Carbonyl Iron Loss Versus Frequency.

12

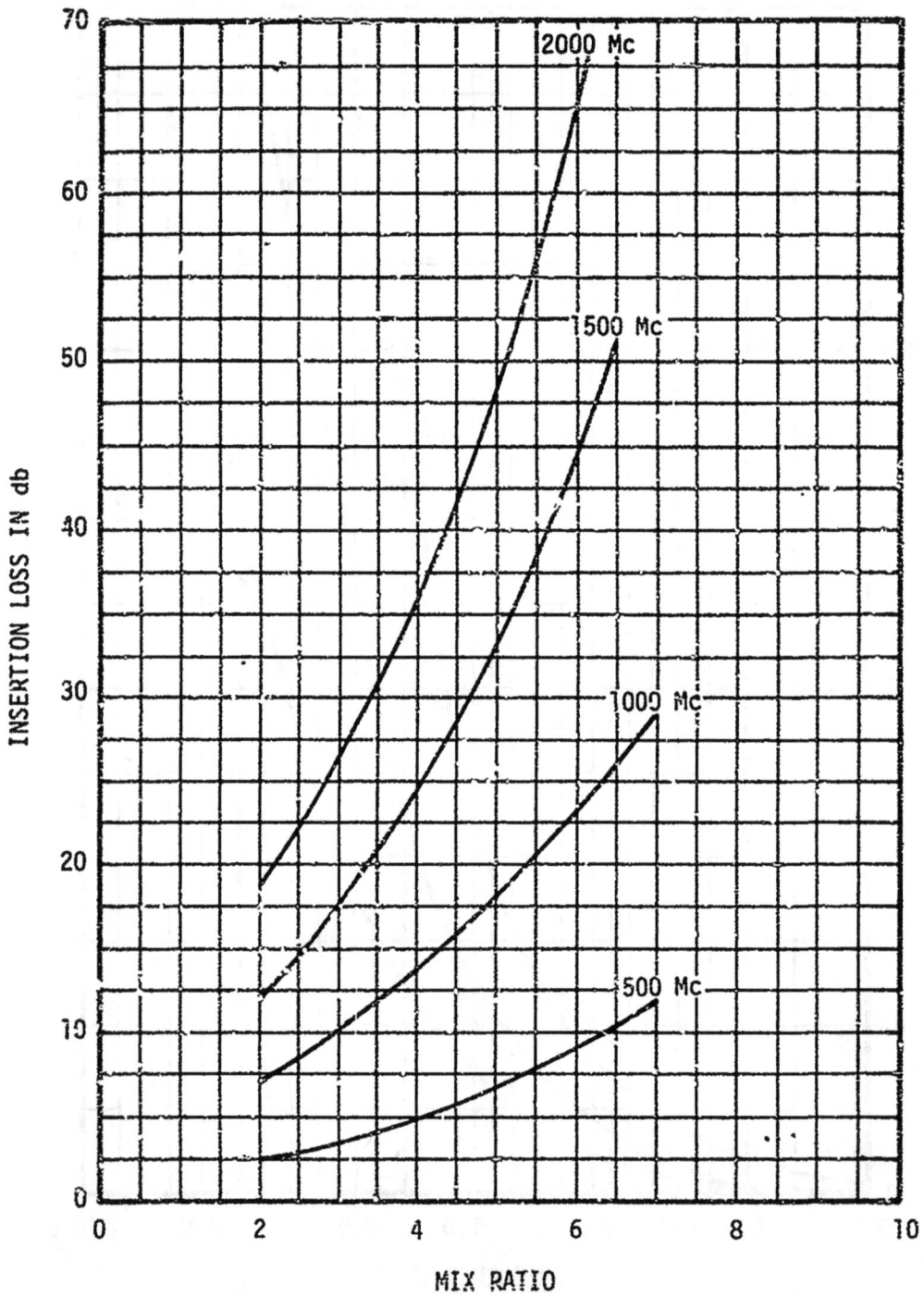

Figure 4. Insertion Loss as a Function of Mix Ratio.

material reduces the attenuation at lower frequencies for a given mix ratio, but increases the attenuation slope somewhat at frequencies where the attenuation is high. Although this steeper attenuation slope is a desirable property, the slight improvement gained by use of the smaller particles is outweighed in most situations by the considerably higher cost of the SF and TH grades of carbonyl iron. Attenuation curves for 6:1 "E" and 6:1 TH test sections are shown in Figure 5 to illustrate the effect of the smaller TH grade particles on attenuation.

2.1.1.2 Ferrites

Attenuation measurements were made on lossy dielectric materials prepared with different grades of ferrite powders. These grades were General Ceramics, Inc. ferrite powders T-1, O-3, H, Q-1 and Q-3. Each powder was formed in a 4:1 ratio by weight with epoxy and used to fill the dielectric space of a short coaxial line section. The results of attenuation measurements made on these line sections are shown in the curves of Figure 6. Generally, the curves exhibit the same general behavior as that of the carbonyl iron, but with two major exceptions. First, large attenuations are encountered at much lower frequencies, indicating that the powdered ferrite materials are suitable for the construction of lossy sections at much lower frequencies than is practical with carbonyl iron. The second distinguishing feature of the curves is the peak of attenuation that occurs in the neighborhood of 6 Gc.

14

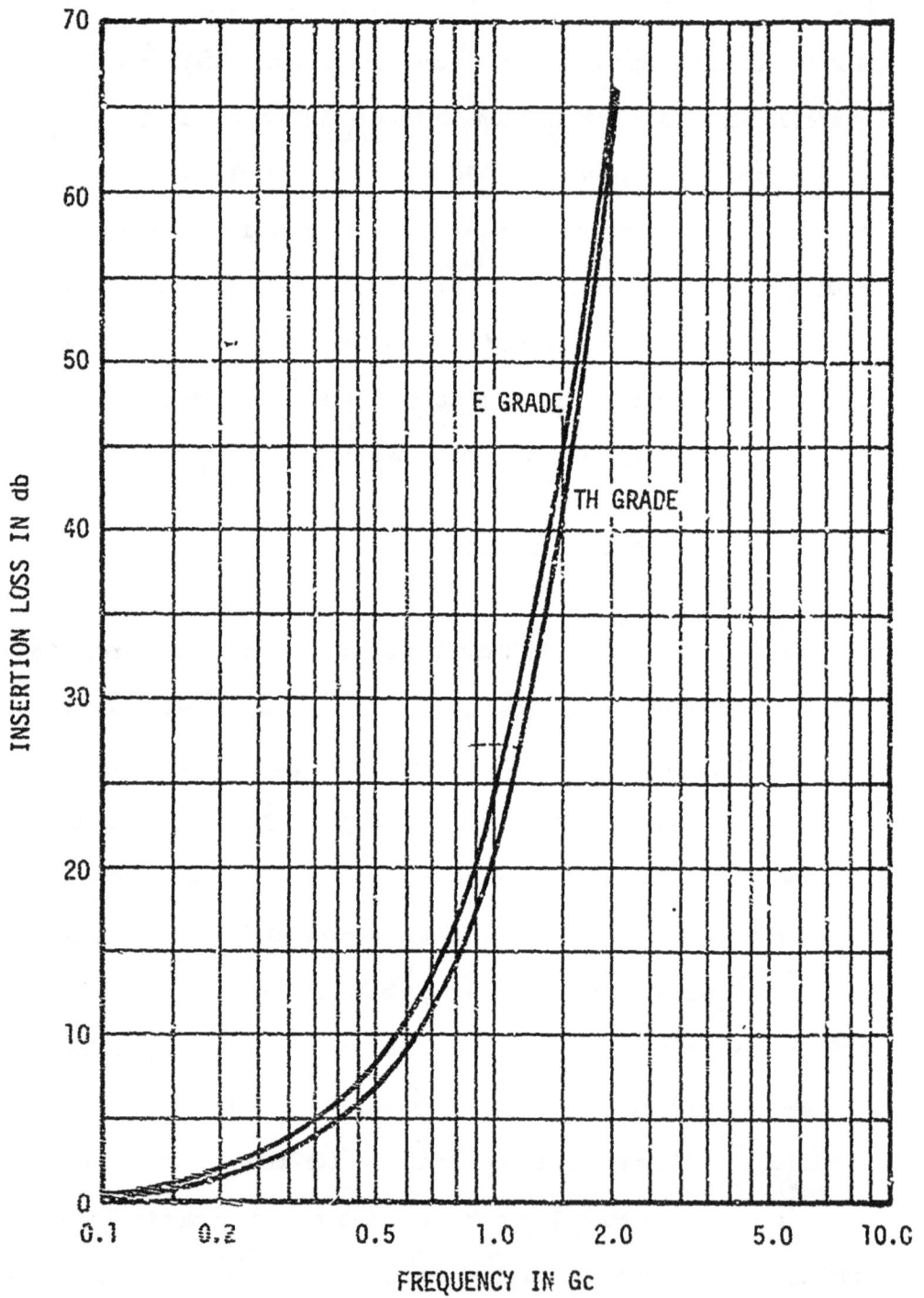

Figure 5. Loss of Two Grades of Carbonyl Iron.

15

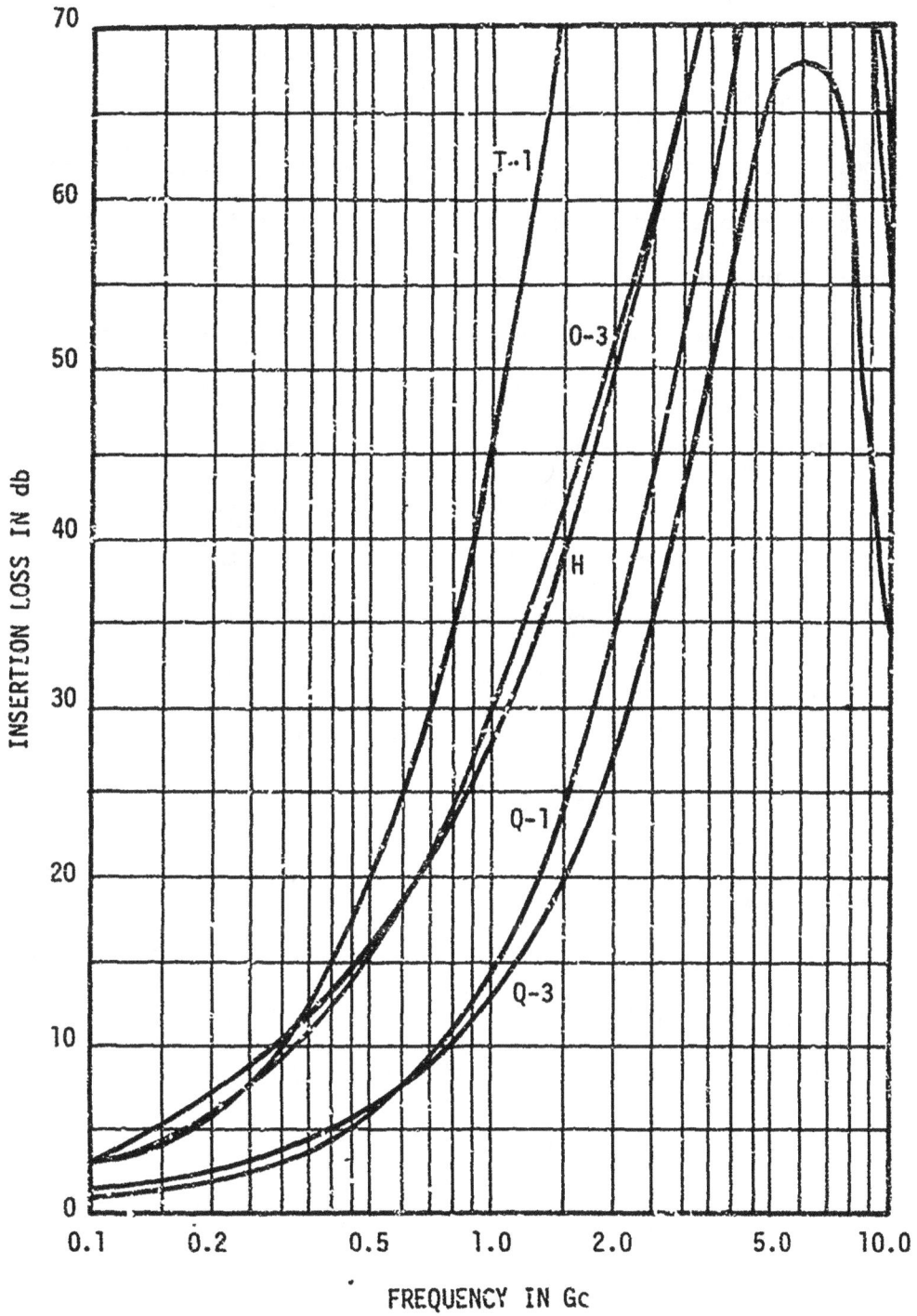

Figure 6. Ferrite Loss Characteristics.

16

Though the full peak is shown only for the Q-3 curve, the low values of attenuation of O-3 and H materials at frequencies between 9 Gc and 10 Gc indicate that a peak of attenuation for both materials occurs in the region near 6 Gc. The data is insufficient to determine if such a peak exists for T-1 and Q-1 materials because the high values of attenuation exceed the sensitivity of the measuring equipment. Shorter sections were constructed using the ferrite materials so that the overall attenuation was lower and the presence or absence of the attenuation peak could be determined. The data obtained on the shorter samples verified the existence of a single attenuation peak for the O-3 and H materials in the vicinity of 6 Gc with the peak for the T-1 material occurring above 10 Gc. Fortunately, the attenuation at 10 Gc is still quite high for the ferrite filled lines and no serious degradation in the ability of ferrite filters to suppress spurious signals would be expected. If for some reason the attenuation at frequencies above 6 Gc is not sufficient to obtain the necessary suppression of spurious signals, a short section of carbonyl iron dielectric material placed in cascade with the ferrite section could provide the necessary increase in attenuation above 6 Gc without adding any appreciable insertion loss at low frequencies.

2.1.1.3 Mill Scale

Another material tested for use as a lossy dielectric was an iron oxide formed in the processing of wrought iron. This material is called mill scale and was supplied by the A. M. Byers

17

Steel Company. In powdered form, mill scale is nonconductive and has
a low frequency dielectric constant of approximately 16 and a
relative permeability of 2. The curve of Figure 7 summarizes the
attenuation data taken on a dielectric material formed with a 4:1
ratio of mill scale to epoxy. The curve shows a fairly steep tran-
sistion from low loss to high loss and relatively low losses at
frequencies under 500 Mc. The original mill scale powder as
received from the H. M. Byers Company contained a wide range of
particle sizes. The distribution of particle sizes was determined
by passing the powder through a series of sieves. The results of
this screening process are given in Table 2. When these data are
plotted on the standard probability scale of Figure 8, the resulting
straight line plot indicates a normal distribution of particle sizes.

TABLE 2

RESULTS OF SCREENING OF A

LARGE QUANTITY OF MILL SCALE

Sample No.	Sample Weight	Size Range
1	247 gms	$d < 44 \mu$
2	561 gms	$149 \mu > d > 44 \mu$
3	1341 gms	$500 \mu > d > 44 \mu$
4	418 gms	$840 \mu > d > 500 \mu$
5	228 gms	$2000 \mu > d > 500 \mu$

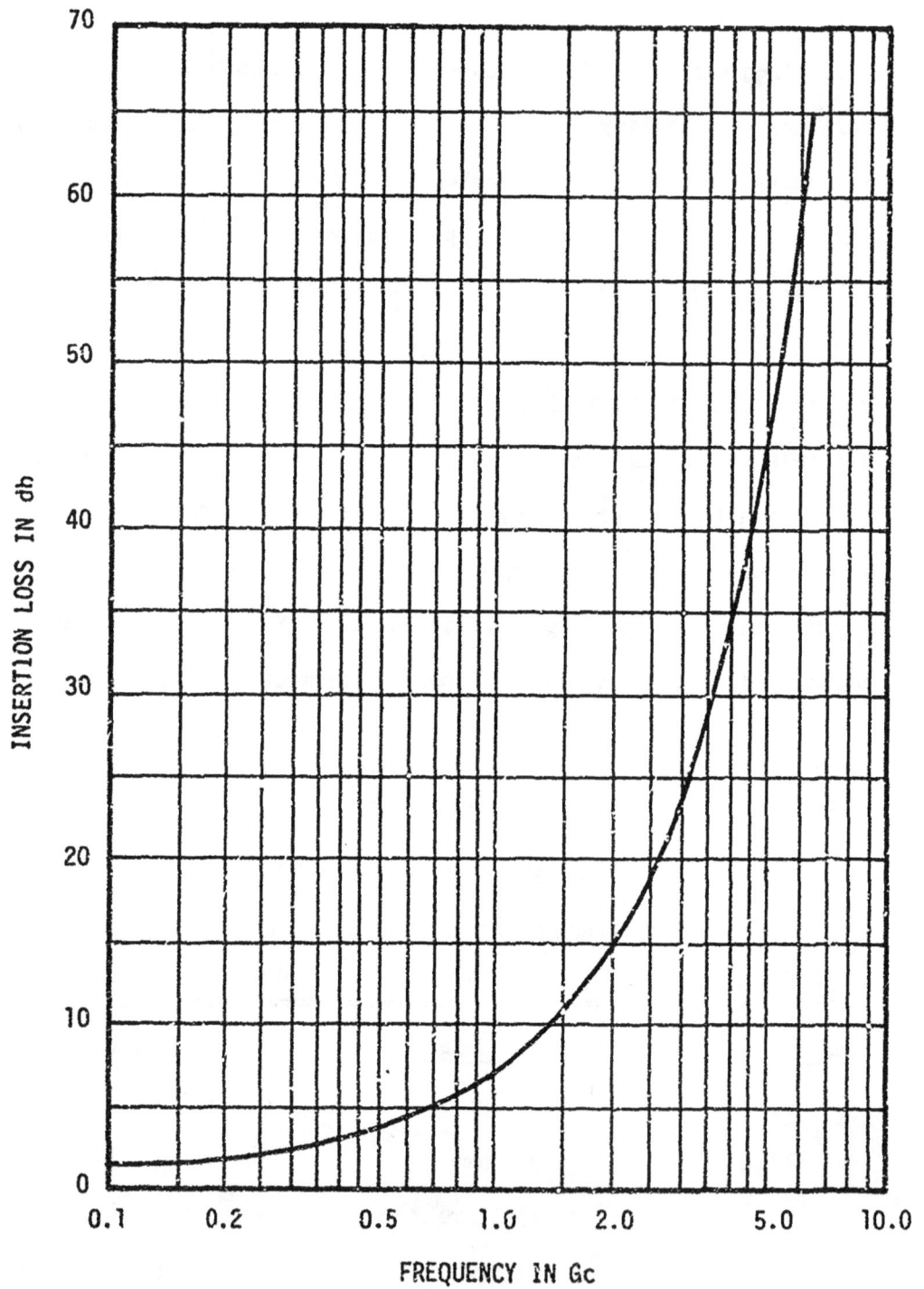

Figure 7. Mill Scale Loss Versus Frequency.

19

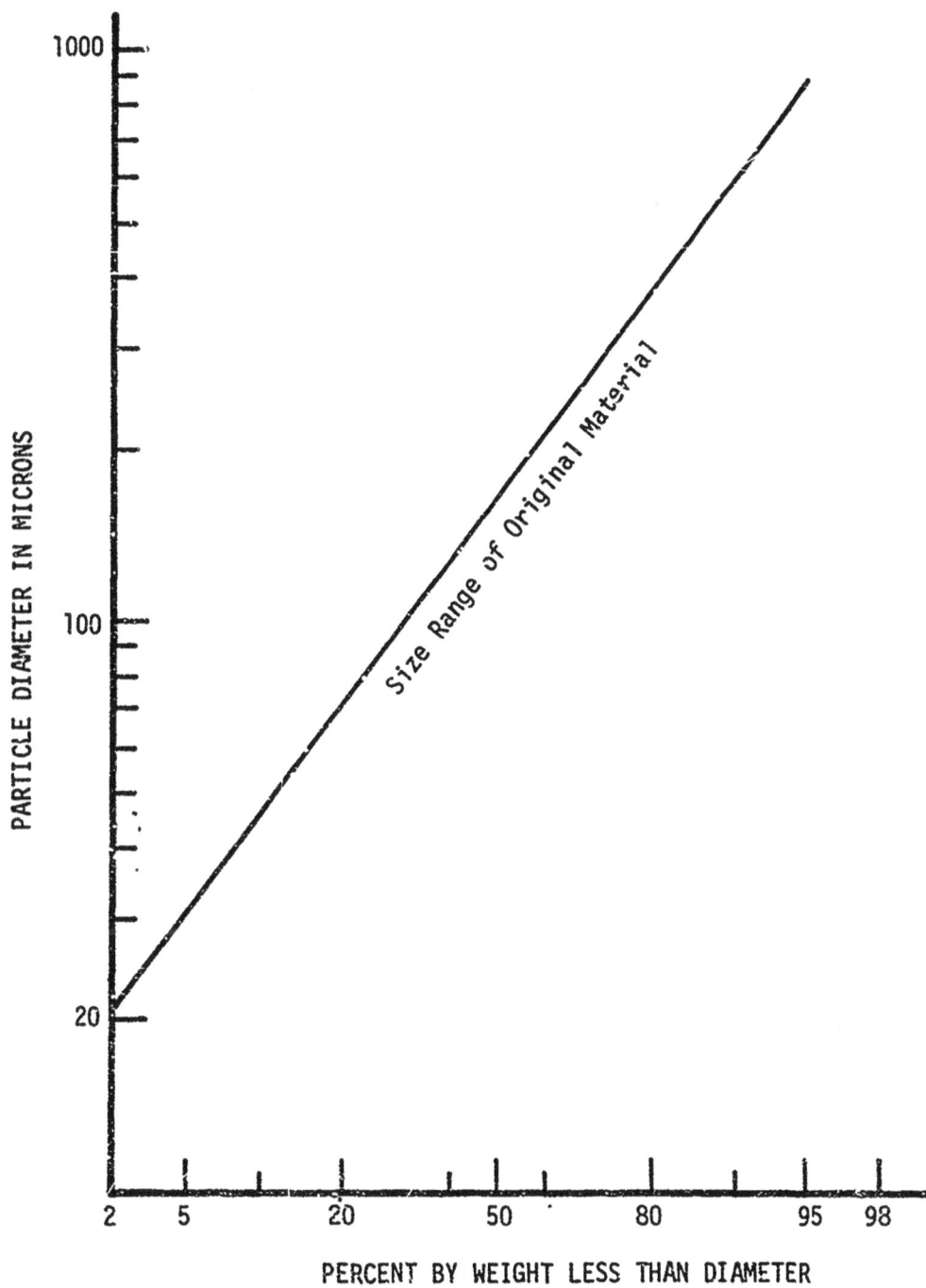

Figure 8. Size Distribution of Mill Scale Particles.

In order to evaluate the effects of particle size on attenuation, test samples of lossy mill scale material were prepared for each of the particle size ranges shown in Table 2, and insertion loss measurements were made on each of the samples. A sample was also prepared using mill scale with particle diameters in the range of 0 to 10 microns. The small particle material was produced by tumbling a portion of the sieved material in a ball-mill. The resulting loss characteristics for six different particle sizes are shown in Figures 9 and 10.

2.1.1.4 Non-Ferrous Powders

Lossy line sections were constructed using several non-ferrous materials in an effort to uncover other materials which might have loss characteristics superior to those measured for the carbonyl iron. The materials tested were cobalt, nickel, aluminum, copper and titanium dioxide. All of these materials were obtained in finely powdered form, prepared in a 2:1 ratio by weight of material to epoxy, and suspended in the epoxy binder in the same fashion as that previously used for the other dielectric materials. Typical loss curves for the materials tested as well as a loss curve for a 2:1 mix ratio of carbonyl iron to epoxy are shown in Figures 11 and 12. The loss curves indicate that the iron material has a low-pass loss characteristic which is superior to all the other materials tested. Not only is the cutoff attenuation slope of the iron higher, but the low frequency attenuation is lower.

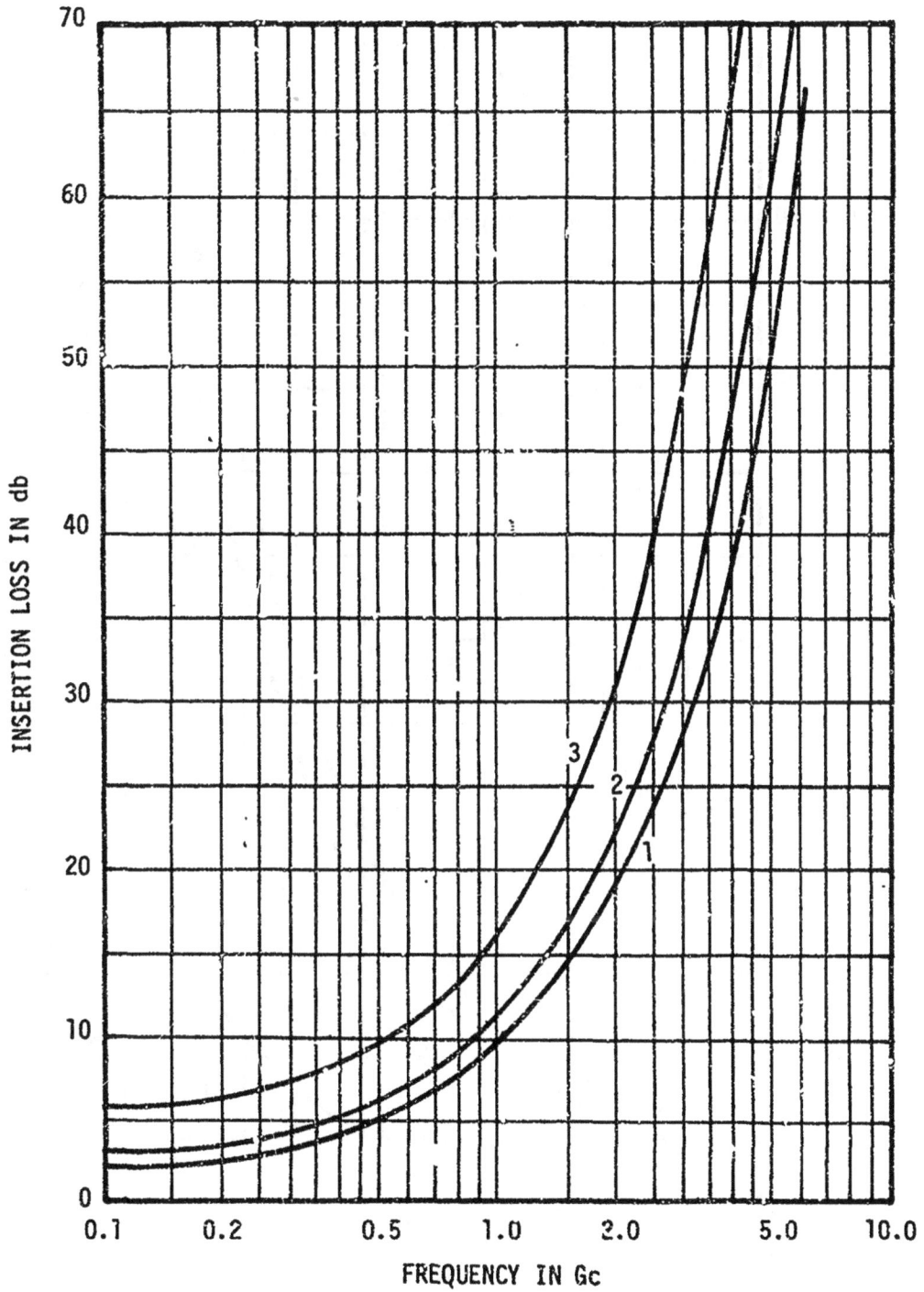

Figure 9.   Effect of Particle Size on Loss.

22

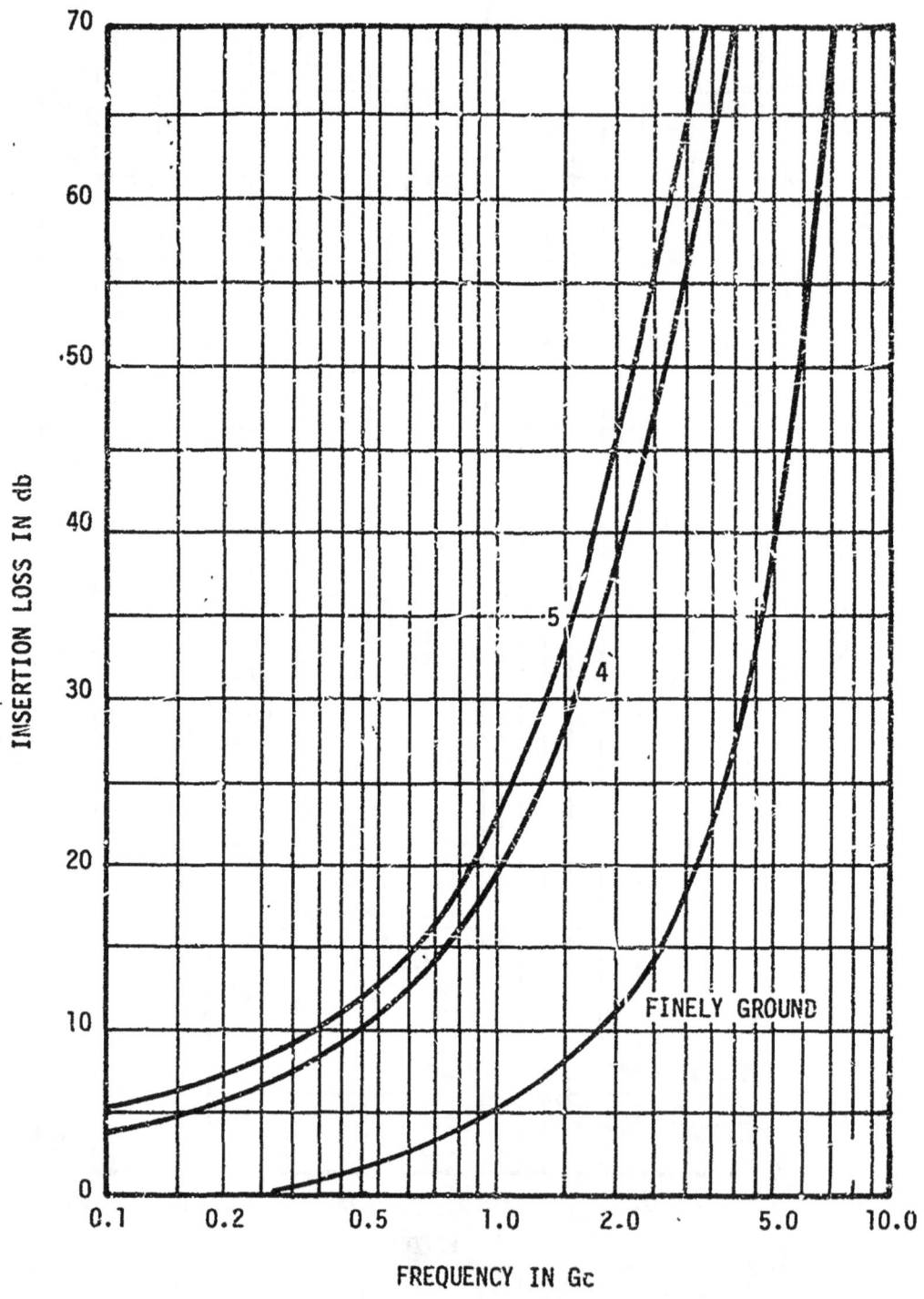

Figure 10. Effect of Particle Size on Loss.

23

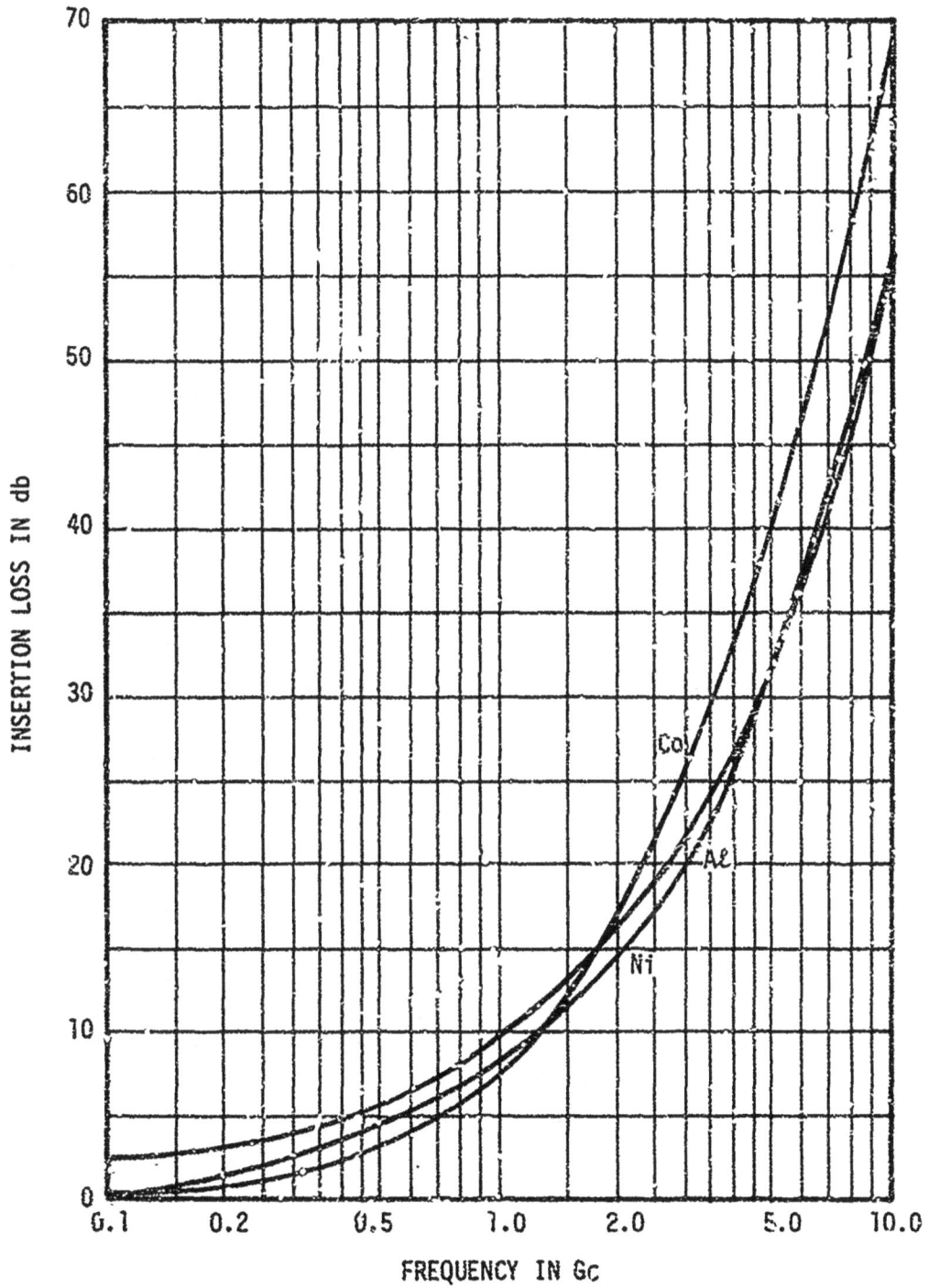

Figure 11. Loss of Non-Ferrous Materials.

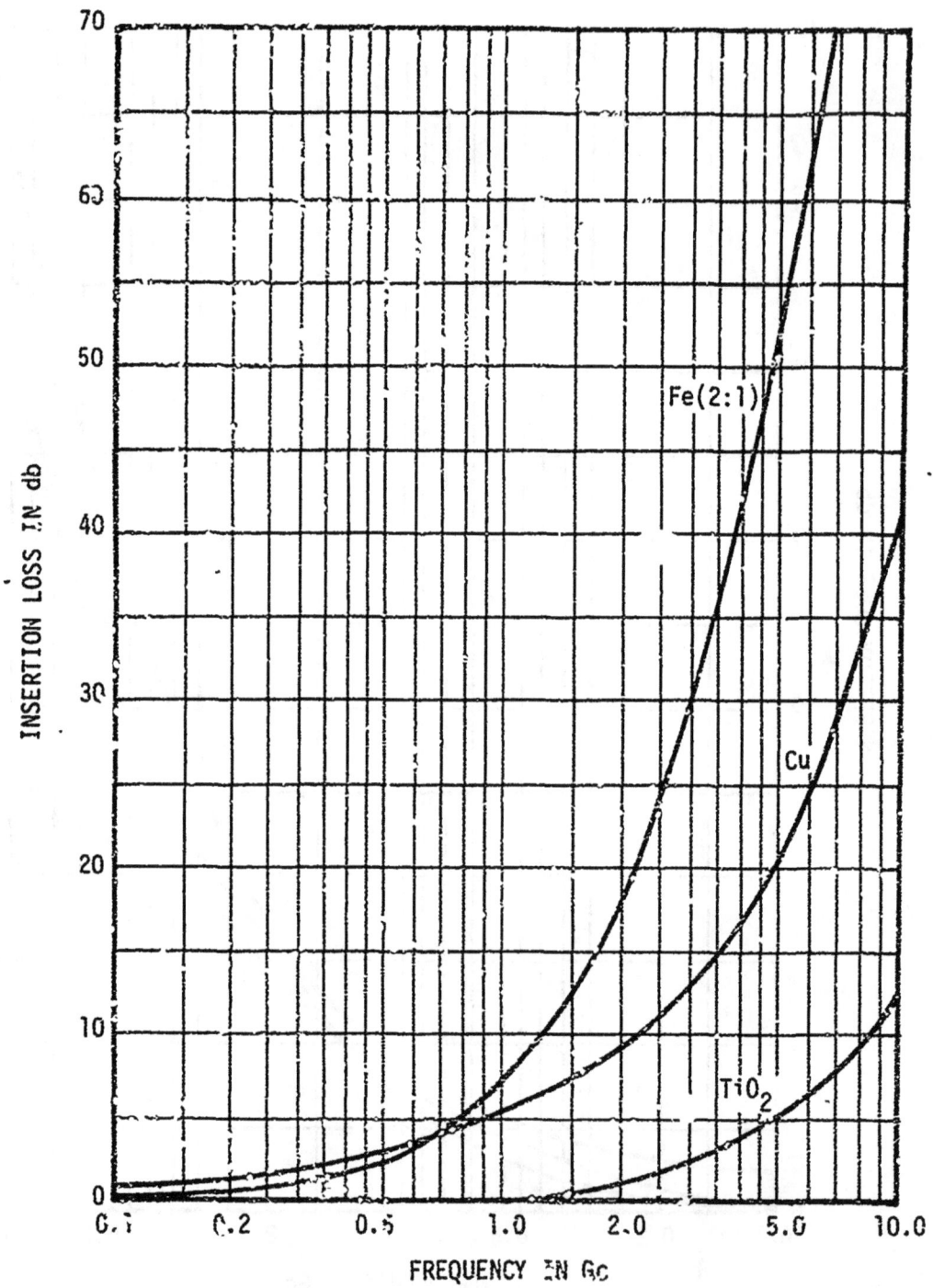

Figure 12. Loss of Non-Ferrous Materials.

The three magnetic materials tested show a direct dependence of the attenuation slope on the permeability of the material indicating that the steeper loss characteristics may be attributed to the additional magnetic losses.

2.1.1.5 Other Ferrous Materials

Nine different ferrous materials obtained from the Glidden Company Metals Division, were tested for application as lossy dielectric materials. These powdered materials included three grades of reduced iron oxide and three grades of electrolytic iron as well as one grade each of carbonyl iron, magnetite and a special blend known by the Glidden identification number D-290. The attenuation versus frequency characteristics of these materials are given in Figures 13 through 17. Several of the materials have characteristics similar to those of the carbonyl iron and the ferrite materials. For example, the loss of Glidden material number D-290 is quite similar to that of carbonyl iron. Also the loss of Glidden material number M-180 closely matches that of the "H" grade ferrite. In both cases, it appears that the two Glidden materials should make satisfactory replacements for the carbonyl iron or ferrite materials.

2.1.2 Measurement of Permeability and Permittivity

A knowledge of the parameters, μ_r^* and ϵ_r^* of the lossy dielectric as necessary to properly select the dimensions of a transmission line filter. An accurate knowledge of these parameters

26

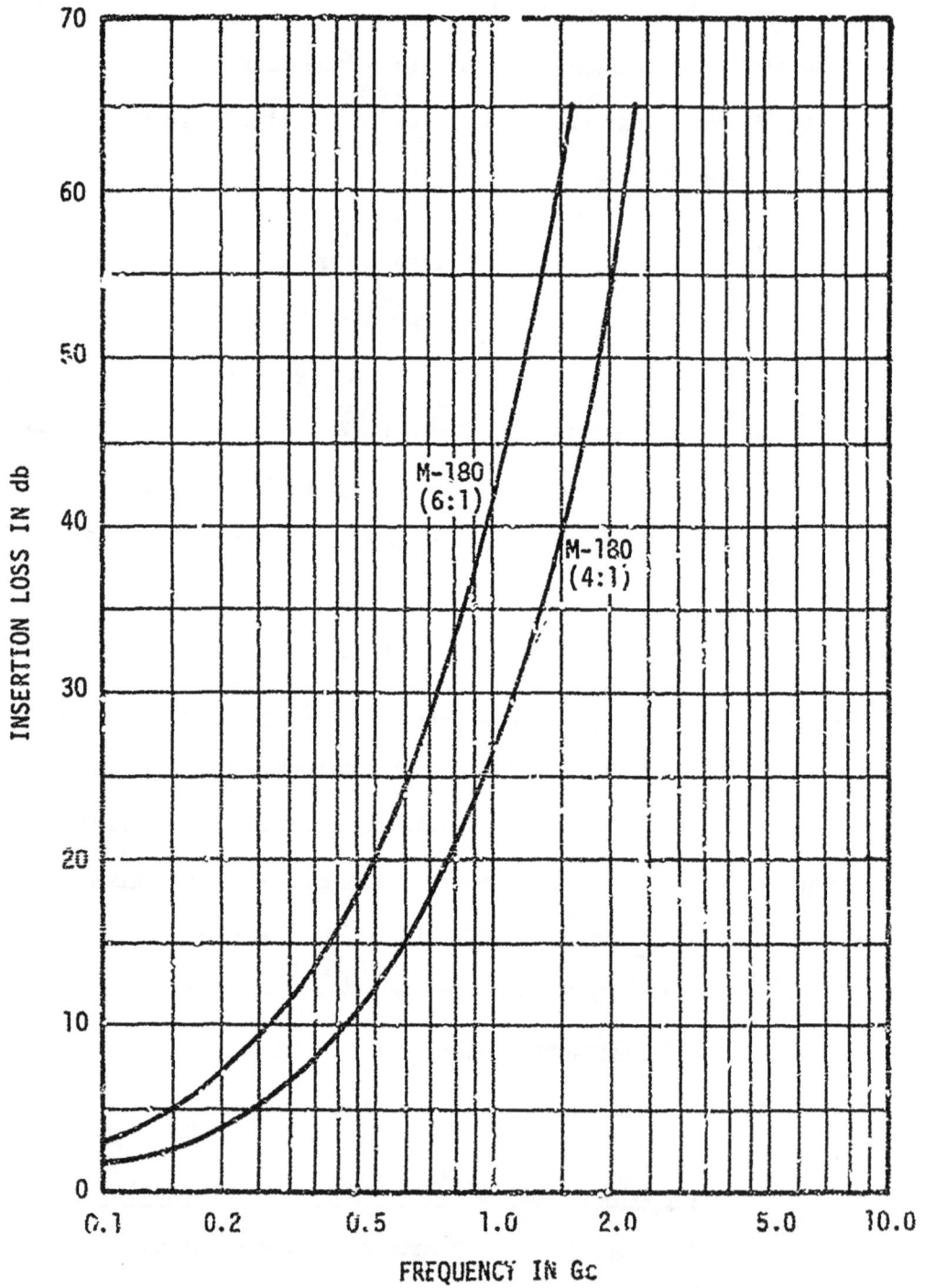

Figure 13. Loss of Magnetite Versus Frequency.

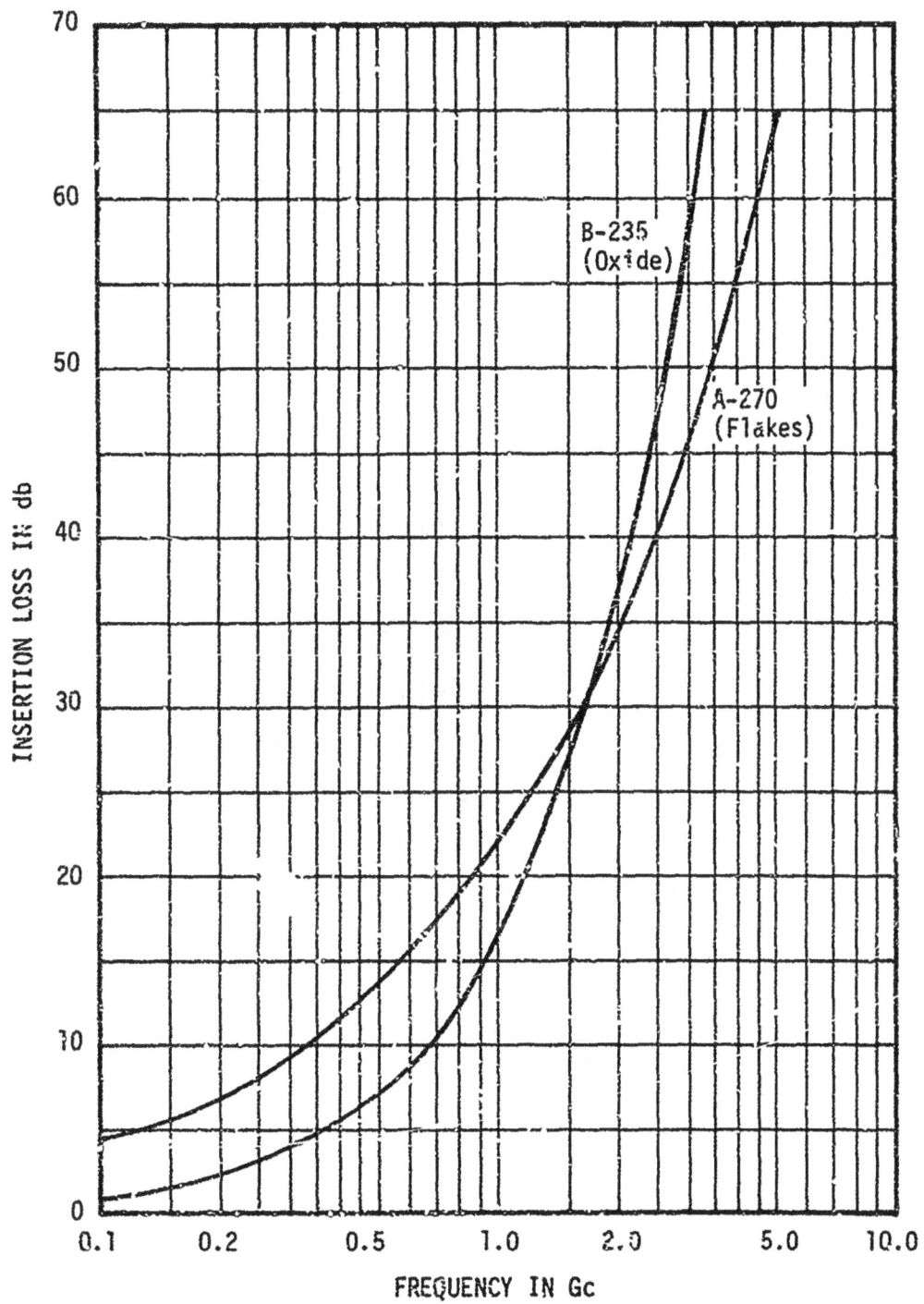

Figure 14. Loss of Iron Oxide and Iron Flakes Versus Frequency.

28

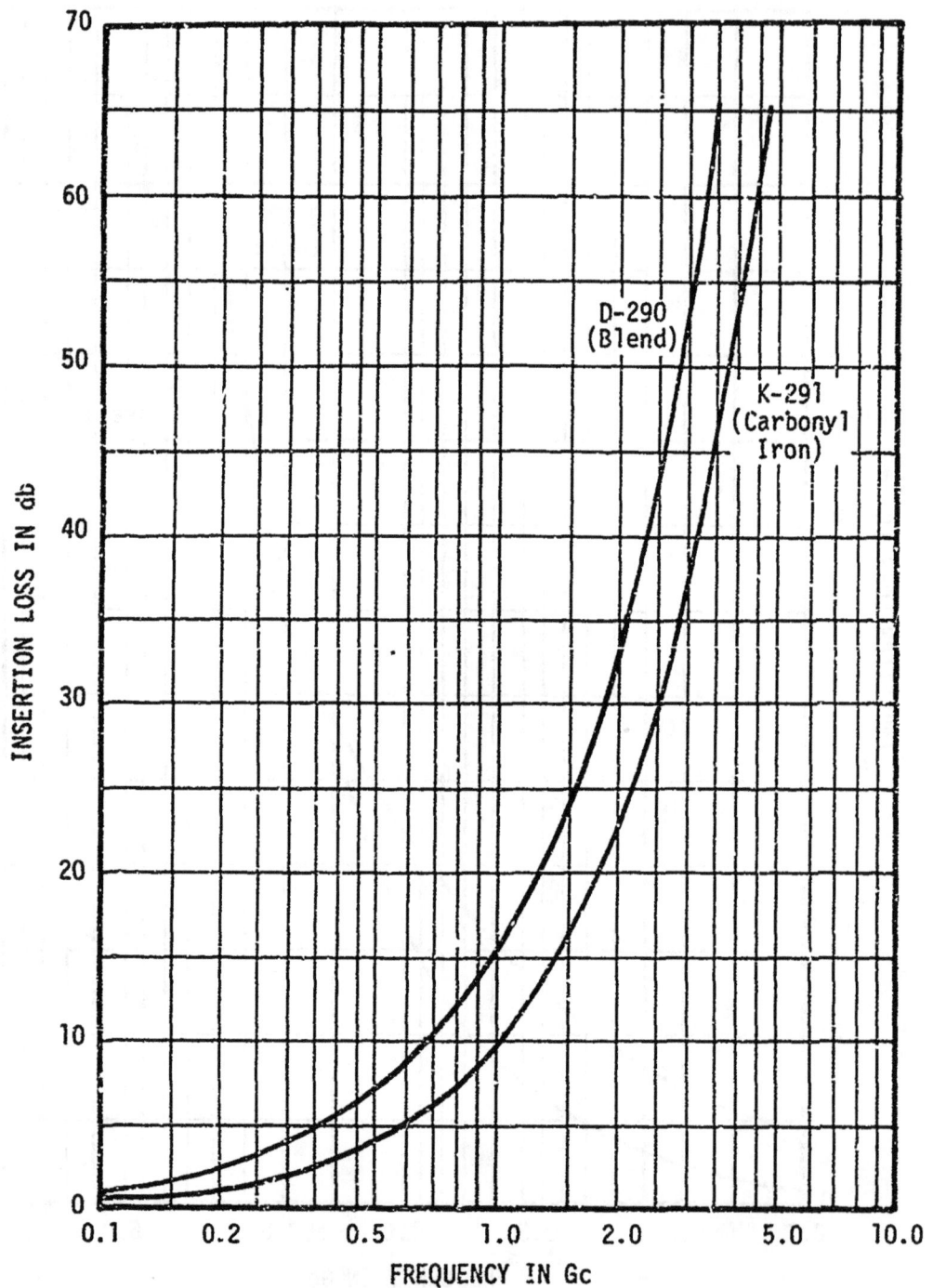

Figure 15.   Loss of Carbonyl Iron and Special Blend Versus Frequency.

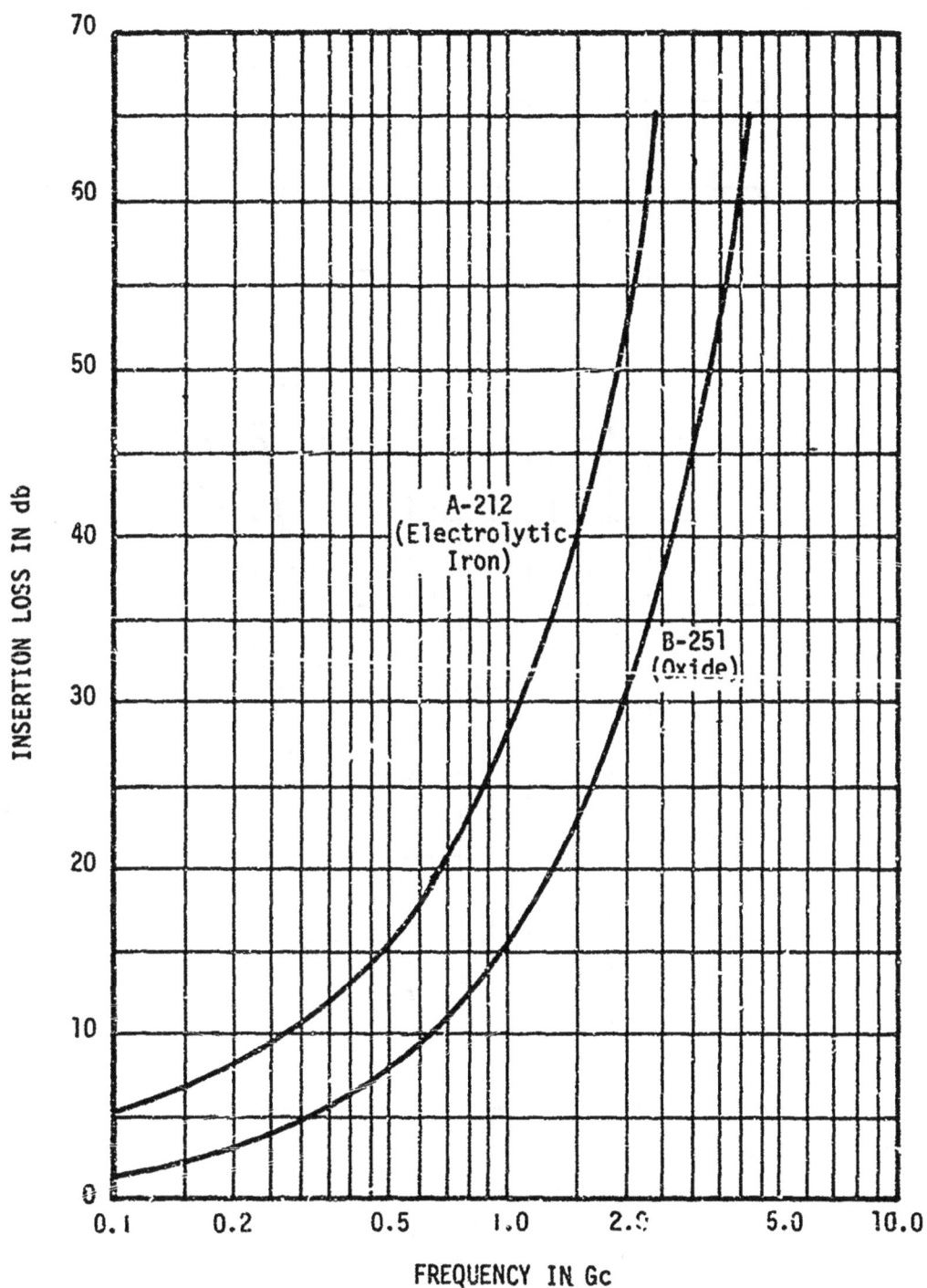

Figure 16.   Loss of Electrolytic Iron and Iron Oxide
Versus Frequency.

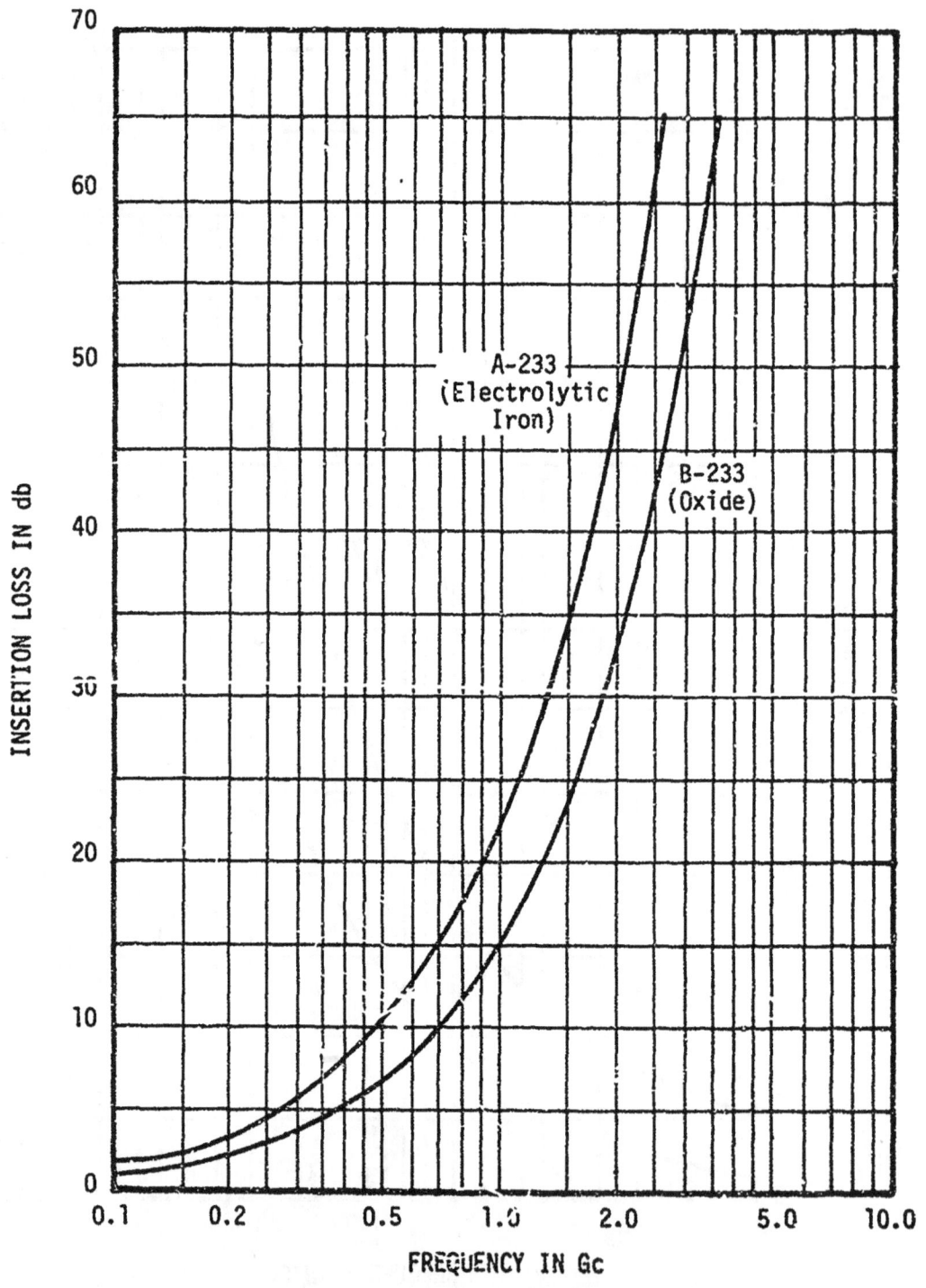

Figure 17. Loss of Electrolytic Iron and Iron Oxide
Versus Frequency.

is most important at frequencies in the filter passband if the passband insertion loss is to be minimized. On the other hand, detailed knowledge of μ_r^* and ϵ_r^* at frequencies in the filter stop-band is of lesser importance since the mismatch losses represent only a small fraction of the total stop-band loss. Since the pass-band regions of the filters considered were generally in the region below 1 Gc, the measurements of μ_r^* and ϵ_r^* were restricted in most instances to the region below 2 Gc, although some of the carbonyl iron mixtures were measured to 4 Gc.

An examination of the literature of μ_r^* and ϵ_r^* measurements indicated that the thin sample technique is widely used for micro-wave measurements. In this technique, an electrically thin sample of material is placed in a transmission line; first in a region of predominantly electric field, and then in a region of predominantly magnetic field. The sample is positioned in a high electric field region by terminating the sample with an open circuit. Practically, this is accomplished by positioning a short circuit $\lambda/4$ from the sample. The predominantly magnetic field condition at the sample position is obtained by terminating the sample with a short circuit. The input impedance of the sample-filled portion of the transmission line is measured for both the open and short circuit conditions. From the impedance data, μ_r^* and ϵ_r^*, can be calculated using the intermediate relations

$$Z_c = \sqrt{Z_{sc} \, Z_{oc}} \tag{11}$$

32

and

$$\tanh \gamma \mathit{l} = \sqrt{\frac{Z_{sc}}{Z_{oc}}} \quad , \tag{12}$$

Z_c is the normalized characteristic impedance, with respect to 50 ohms, of that portion of the coaxial transmission line containing the material under test. The quantities, Z_{sc} and Z_{oc} are the normalized short and open circuit input impedances existing at the face of the test sample. The propagation constant of the section of line containing the dielectric material, is γ and l is the length of the sample. The complex relative permeability, μ_r^*, and relative dielectric constant, ϵ_r^*, are evaluated from the relationships

$$\epsilon_r^* = - j \frac{\gamma \lambda_0}{2\pi} \cdot \frac{1}{Z_c} \quad , \tag{13}$$

and

$$\mu_r^* = - j \frac{\gamma \lambda_0}{2\pi} \cdot Z_c \quad , \tag{14}$$

where λ_0 = free space wavelength.

The thin sample technique assumes that measurements are made of a dielectric sample whose thickness is much less than the electrical wavelength inside the material. It is difficult to estimate the wavelength inside the material being tested until some knowledge of μ_r^* and ϵ_r^* has been established. The difficulty in estimating the electrical thickness of the sample was avoided by observing

33

the shift in the slotted line null pattern when the sample was inserted. A shift in the null pattern of $\lambda/16$ or less, indicated that the sample was electrically thin.

Some insight into the reasoning behind the $\lambda/16$ criterion can be gained from Figure 18. The top of the figure shows a standing wave pattern illustrating the variation of the electric and magnetic fields, with distance along the transmission line. The distance between null points is equal to one-half wavelength. It can be seen that the two one-sixteenth wavelengths that occupy the center region between the nulls of the E field are essentially high E field and low H field regions. The lower portion of the figure shows the standing wave pattern (dotted curve) that exists when the sample is positioned $\lambda/4$ from the shorted end of the transmission line. The distance $\underline{d}$ between the null points of the two standing wave patterns indicates the effect of the sample. If the distance $\underline{d}$ does not exceed $\lambda/16$, then the portion of the standing wave pattern that exists inside the sample is less than $\lambda/16$, and the sample remains in an essentially constant E field and an essentially zero H field. Consequently, only the dielectric constant of the sample is effective in producing a shift of the null pattern.

The sample holder shown in Figure 19 was constructed for use with the precision slotted line. This holder was designed with Type N connectors to mate with the slotted line. The movable short is connected at the sample end of the holder and is capable of being positioned as close as desired to the sample. The short is moved by

34

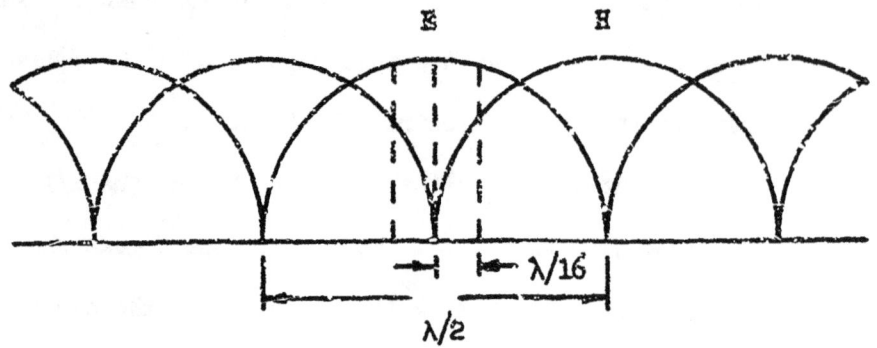

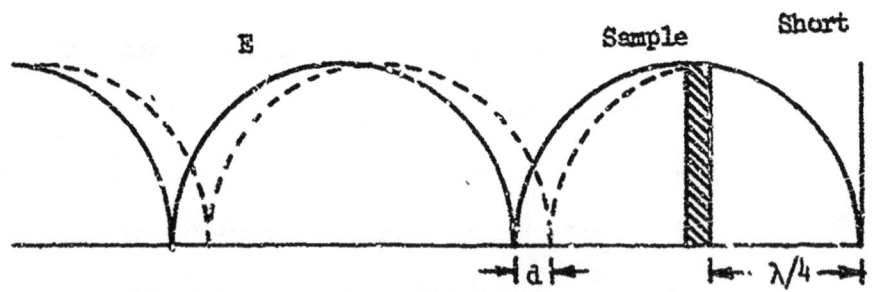

Figure 18. Standing Wave Patterns.

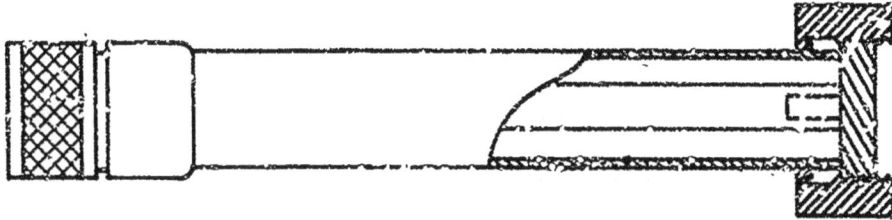

(a) CUTAWAY VIEW OF SAMPLE HOLDER.

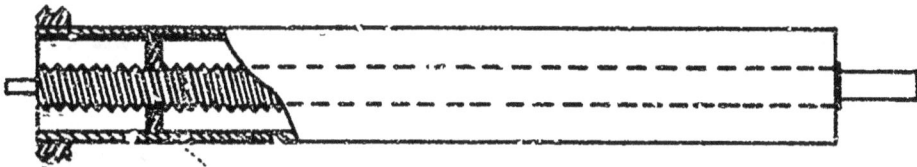

(b) CUTAWAY VIEW OF MOVABLE SHORT.

Figure 19. Sample Holder and Movable Short.

the use of a threaded center conductor which mates with the sample
holder. Good electrical contact with the short circuit is assured
by providing a pin on the shorting plug which fits tightly in the
center conductor, while contact with the outer conductor is assured
by tightening the plug down with a threaded sleeve on the outer
conductor. The threaded sleeve also pulls the movable short tightly
against the sample holder. Values of VSWR for the unloaded holder
in excess of 100 were measured at frequencies up to 4 Gc, indicating
that the holder had low losses.

Over 400 different measurements of μ_r^* and ϵ_r^* were made using
the equipment setup shown in Figure 20. Representative results of
these tests are presented in Tables 3 through 6. Some of the data
obtained on μ_r^* and ϵ_r^* was erratic. While some of the data points
could be eliminated on the basis of errors in the data acquisition,
other variations could not be easily explained. At practically every
test frequency, at least two and, when available, three sample
thicknesses were measured, but close agreement between the samples
was not always obtained. A particular sample thickness did not
always behave uniformly with frequency. Many of the measurements
were repeated at widely separated times to evaluate the results of
measurement variations. Generally, normal measurement error did
not explain all the variations observed. The most consistent results
were obtained with the measurements of the carbonyl iron samples.
Figure 21 shows the characteristic behavior of μ_r^* and ϵ_r^* of carbonyl
iron up to 2 Gc.

37

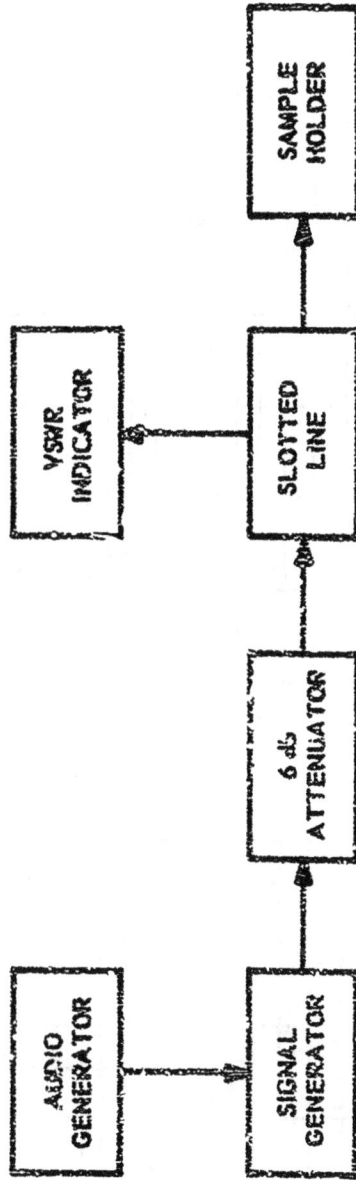

Figure 20. Equipment Arrangement for μ_r^* and ε_r^* Determination.

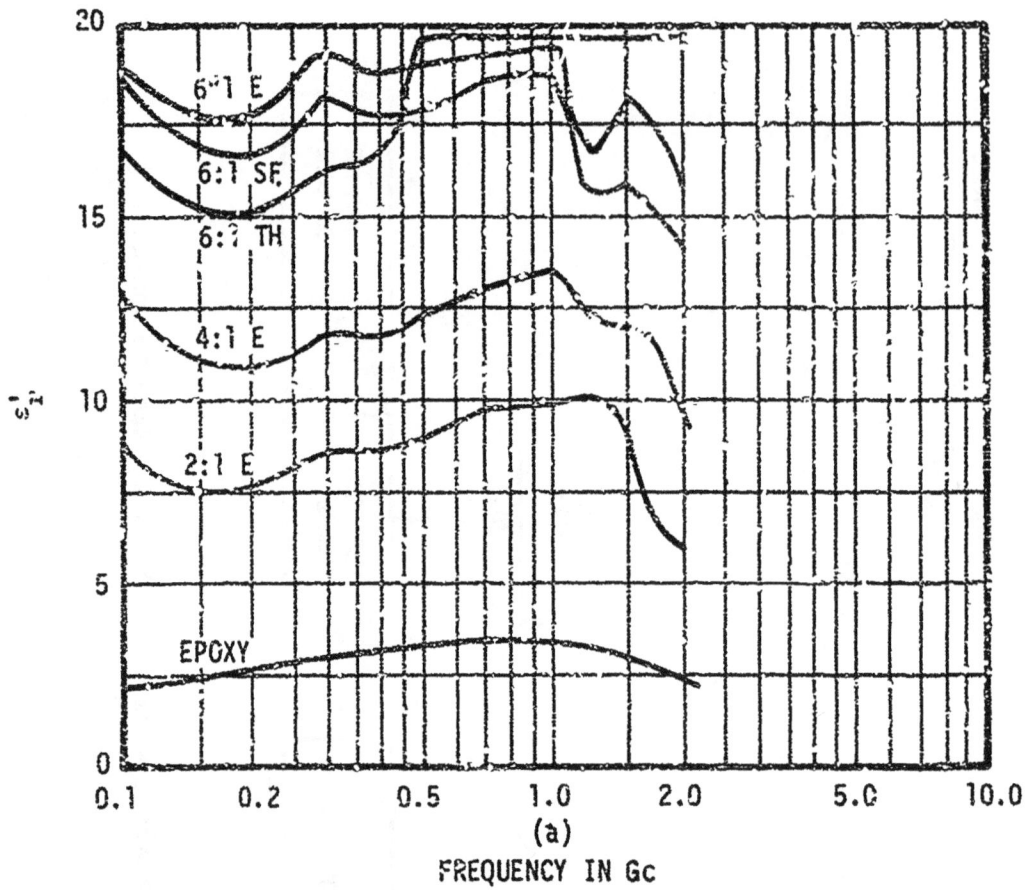

(a)

FREQUENCY IN Gc

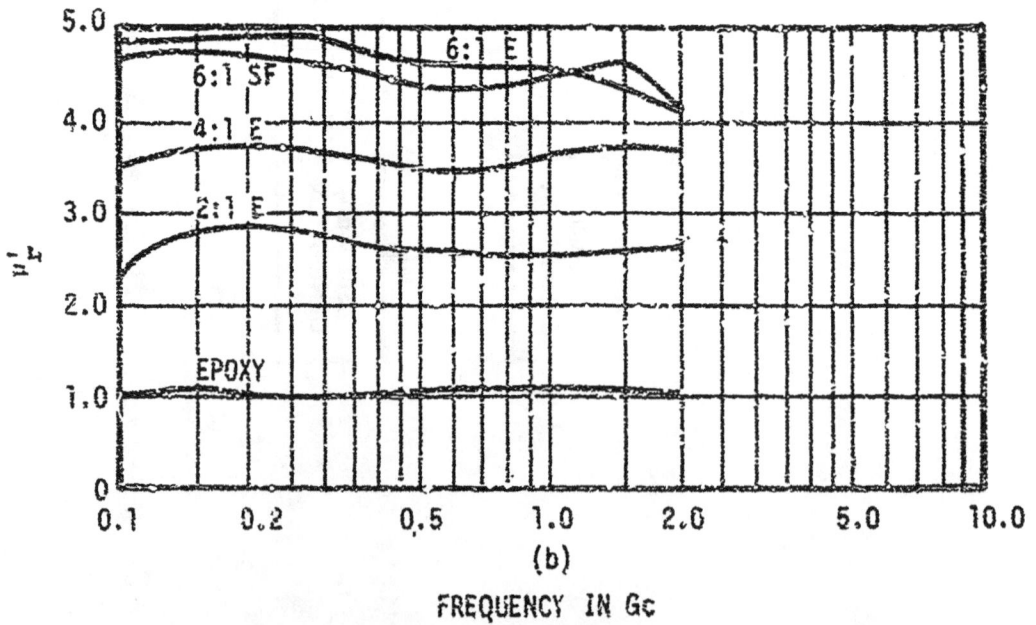

(b)

FREQUENCY IN Gc

Figure 21. μ_r' and ϵ_r' of Iron Dielectric Materials.

The data shows that the dielectric constant values have a greater variation than the permeabilities. Since the measured value of the dielectric constant is more sensitive to small gaps between the sample and its holder than the relative permeability, the effect of errors in sample positioning would be expected to be greater on the dielectric constant measurements than on the permeability measuremer .

Another source of error is the violation of the thin sample criterion. The parameters of the absorbing mixtures make the electrical thickness much greater than the physical thickness and it is the electrical thickness which must be thin with respect to the wavelength. At the higher test frequencies, the physical thickness of the samples must be very small, which creates two additional problems. One is the lack of physical strength of the sample which causes both machining and handling to be difficult without breakages. The other problem is related to the technique used to make the samples. Bubbles generated in the mixing of the lossy powder and epoxy are difficult to completely remove from the material before the hardening process begins so that small voids are created in the dielectric material. The effect of these voids becomes significant in very thin samples where they may result in holes which extend completely through the sample.

The reduction of the raw data taken with a slotted line or an impedance bridge to produce values of μ^* and ϵ^* is a time consuming process which is subject to numerical error. In order to speed up this computation process, a computer program was written to produce values of μ_r^* and ϵ_r^* from the impedance data. The program evaluated Equations (11) and (12) in terms of the slotted line data. These results were then substituted in Equations (13) and (14) to produce the printout in terms of μ_r', μ_r'', ϵ_r', and ϵ_r''.

TABLE 3

PERMEABILITY AND PERMITTIVITY OF CARBONYL IRON MATERIALS

2:1 E	μ_r'	μ_r''	ϵ_r'	ϵ_r''
100 Mc	2.58	0.00	8.85	0.66
200	2.95	0.00	7.55	0.60
300	2.78	0.15	8.73	0.51
400	2.57	0.10	8.67	0.46
500	2.63	0.23	9.12	0.48
700	2.49	0.32	9.70	0.49
1000	2.45	0.43	9.89	0.38
1200	2.70	0.52	10.22	0.36
1500	2.62	0.69	9.11	0.28
1700	2.58	0.94	6.52	0.20
2000	2.60	0.76	5.99	0.31
4:1 E				
100 Mc	3.56	0.00	12.81	1.12
200	3.73	0.01	10.87	1.00
300	3.71	0.26	11.87	0.82
400	3.53	0.27	11.60	0.76
500	3.51	0.43	12.27	0.84
700	3.34	0.52	12.96	0.87

TABLE 3 (Continued)

1000	4.05	0.83	13.66	0.68
1200	3.65	1.10	12.39	0.45
1500	4.25	1.35	13.34	0.31
1700	4.00	1.50	11.58	0.49
2000	3.51	1.26	9.54	0.69

6:1 E

100 Mc	4.93	0.00	19.17	0.96
200	5.21	0.00	16.77	0.00
300	5.00	0.46	19.48	0.88
400	4.74	0.54	18.97	0.71
500	4.69	0.71	19.26	0.93
600	4.63	0.83	19.32	0.93
700	4.60	0.93	19.47	0.93
800	4.52	1.05	19.29	0.88
1000	4.53	1.10	19.71	1.53
1200	4.24	2.74	15.56	2.67
1500	4.45	2.08	15.85	0.25
1700	4.44	2.25	15.02	0.05
2000	4.14	2.17	14.46	0.31

6:1 SF

100	4.71	0.00	18.78	1.11
200	4.78	0.01	16.51	0.86
300	4.71	0.26	18.22	1.09
400	4.58	0.33	17.55	1.03
500	4.58	0.37	17.94	1.09
600	4.38	0.52	18.24	1.03
700	4.42	0.61	18.60	1.09
800	4.35	0.73	18.58	1.05
1000	4.45	0.94	18.81	1.29
1200	4.60	1.57	16.47	0.38

TABLE 3 (Continued)

1500	4.68	1.69	18.38	0.96
1700	4.48	1.77	17.41	1.18
2000	4.03	1.64	15.97	1.97

6:1 TH

100 Mc	4.60	0.00	16.88	0.73
200	4.70	0.21	14.74	0.66
300	4.47	0.29	16.37	0.68
400	4.22	0.34	16.29	0.66
500	4.87	0.53	19.82	0.92
600	4.69	0.59	20.24	0.93
700	4.72	0.68	20.64	0.81
800	4.50	0.79	20.55	0.84
1000	4.16	1.07	20.41	0.70

TABLE 4

PERMEABILITY AND PERMITTIVITY OF FERRITE MATERIALS

4:1 0-3	μ_r'	μ_r''	ϵ_r'	ϵ_r''
400 Mc	3.70	1.31	7.97	2.44
600	3.49	1.60	7.44	2.30
800	3.54	1.75	7.48	2.08
1000	2.44	1.79	7.17	2.08
4:1 Q-1				
400 Mc	3.64	1.00	8.17	0.28
600	3.44	1.19	6.58	0.33
800	3.62	1.27	5.40	0.33
1000	2.66	1.61	6.12	0.33

TABLE 4 (Continued)

4:1 Q-3

400 Mc	2.81	0.89	5.93	0.68
600	2.74	1.09	5.27	0.26
800	2.97	1.19	5.37	0.19
1000	2.44	1.29	5.44	0.33

4:1 T-1

400 Mc	4.18	4.01	16.18	1.16
600	4.46	2.41	17.52	1.90
800	3.90	3.18	15.05	1.59
1000	2.81	2.78	16.85	1.84

TABLE 5

PERMEABILITY AND PERMITTIVITY OF MILL SCALE MATERIALS

MS No. 1	μ_r'	μ_r''	ϵ_r'	ϵ_r''
400 Mc	1.92	0.13	15.00	0.83
600	1.54	0.26	14.11	0.74
800	1.58	0.18	15.56	0.78
1000	1.68	0.24	15.16	1.05
MS No. 2				
400 Mc	1.93	0.17	17.50	1.58
600	1.67	0.12	16.59	1.15
800	1.55	0.15	15.52	1.06
1000	1.68	0.37	11.81	0.81
MS No. 3				
400 Mc	2.32	0.21	19.31	0.58

600	2.15	0.06	15.26	1.52
800	2.35	0.20	13.46	1.07
1000	2.38	0.31	17.74	1.52

MS No. 4

400 Mc	1.74	0.32	16.82	1.97
600	1.43	0.35	14.77	1.52
800	1.44	0.30	17.72	1.93
1000	1.68	0.44	16.59	2.03

MS No. 5

400 Mc	1.58	0.42	17.59	1.62
600	1.45	0.43	13.37	1.18
800	1.36	0.47	13.31	1.24
1000	1.50	0.39	14.84	2.92

TABLE 6

PERMEABILITY AND PERMITTIVITY OF NON-FERROUS MATERIALS

Aluminum	μ'_r	$\tan \delta_\mu$	ϵ'_r	$\tan \delta_\epsilon$
1000 Mc			19.6	-0.04
700			20.9	-0.07
500			17.7	-0.03
Copper				
1000 Mc			19.7	-0.03
700			18.9	-0.02
500			17.5	-0.02

TABLE 6 (Continued)

Cobalt

1000 Mc	1.5	-0.19	12.3	-0.04
700	1.5	-0.20	8.9	-0.04
500	1.6	-0.19	10.5	-0.04

Nickel

700 Mc	1.2	-0.33	14.2	-0.03
500	1.3	-0.31	12.3	-0.03

Titanium Dioxide

1000 Mc	6.6	-0.03
700	8.9	-0.01
500	4.5	-0.03

2.1.3 Design of Tapers

Because the μ_r^* and ϵ_r^* of the dielectric material are in a different ratio than that of air, it is necessary to change the dimensions of the coaxial line inside the dielectric filled region to avoid an impedance discontinuity at the boundary between the air and the dielectric material. The proper values for the dimensions of a 50 ohm coaxial line with a dielectric material between the center and outer conductors can be computed from the relation

$$\frac{r_1}{r_2} = e^{\left(\frac{5}{6}\sqrt{\frac{\epsilon_r}{\mu_r}}\right)} , \tag{15}$$

where r_1 and r_2 are the radii of the outer and inner conductors respectively. The values of μ_r and ϵ_r of the carbonyl iron material vary with frequency, and the characteristic impedance of the lossy line section is not constant. The permeability falls more rapidly with increasing frequency than the dielectric constant so that the characteristic impedance falls to low values at very high frequencies.. A typical frequency variation of the input VSWR of a section whose dimensions were chosen to produce a low frequency characteristic impedance of 50 ohms is shown in Figure 22(a). An improvement in the match at high frequencies can be obtained by employing a short linear taper of the dielectric material of the form shown in Figure 22(b) rather than an abrupt change from air to the lossy dielectric. Notice that the VSWR curve of Figure 22(b) shows a considerable improvement at frequencies above 3 Gc for this length of taper. The improvement results from the fact that reflections from the mismatched region where the dielectric completely fills the coaxial line, are attenuated in the short tapered section. However, the low frequency passband has a large VSWR because the diameter of the center conductor in the tapered section is too small to provide a 50 ohm match in the passband region. Some improvement in the low frequency VSWR can be obtained by continuing the center conductor through the taper with the same diameter that was used in the air line. Nevertheless, appreciable low frequency reflections are obtained from the tapered section since the characteristic impedance of this section is continually changing over the length of the taper. These reflections

47

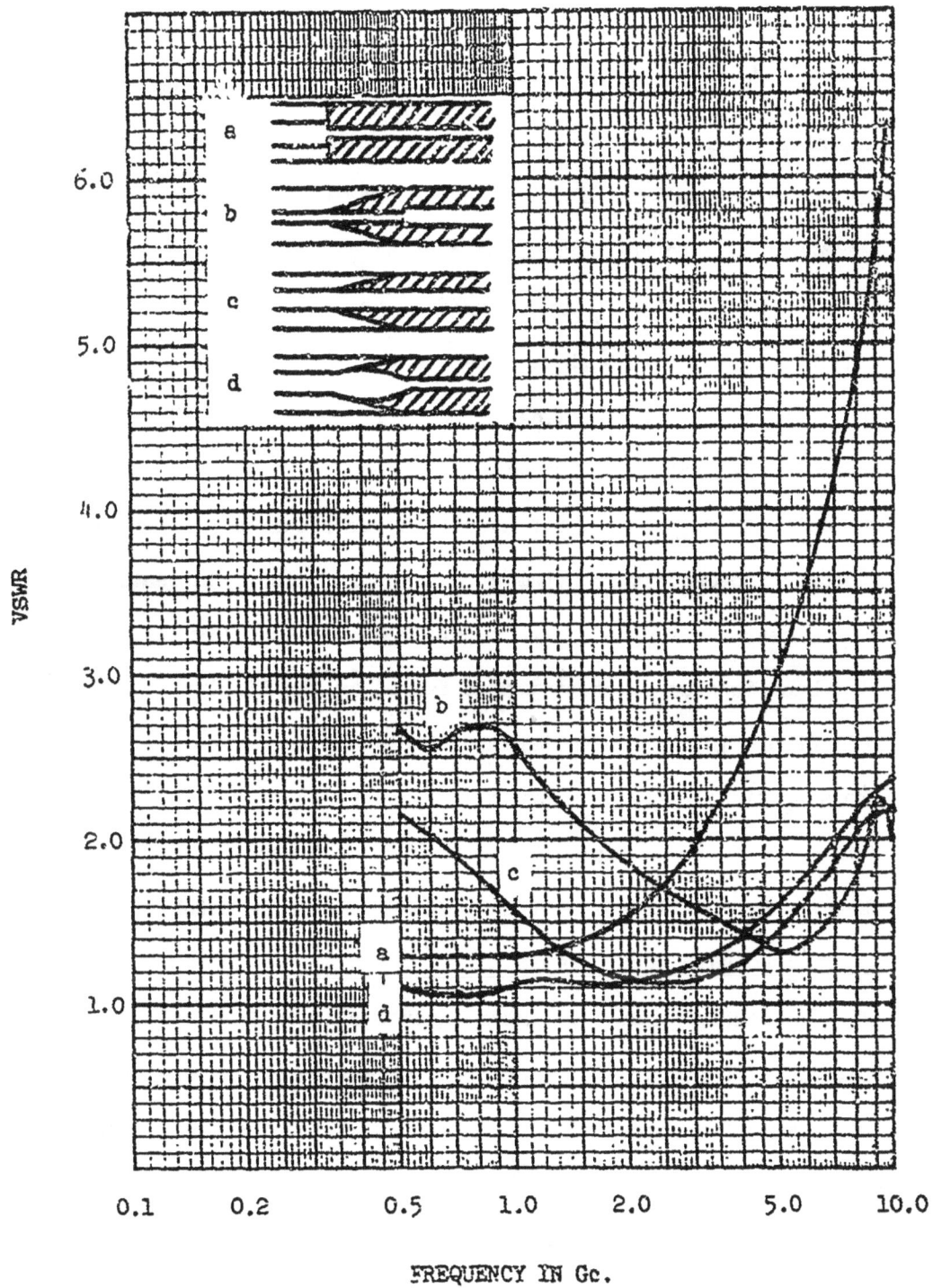

Figure 22. Effect of Taper on VSWR.

48

are not dissipated in the taper because the losses in the dielectric material are low at frequencies in the filter passband. The behavior of the input impedance for the case where the center conductor is continued through the taper with same diameter as was used in the air line is illustrated by the VSWR curve of Figure 22(c). It is possible to maintain both a stop-band and passband match by tapering the diameter of the center conductor over the length of the taper of the dielectric material. Appendix C outlines a procedure for determining the center conductor taper required to provide a passband impedance of 50 ohms at every point along the center conductor over the length of the taper of the dielectric material. Stop-band impedance matching is obtained by the dissipation of reflections in the linear taper of the dielectric material. A cross sectional view of a taper designed by the procedure of Appendix C is shown in Figure 23. The VSWR curve of Figure 22(d) illustrates the considerable improvement in passband impedance match that can be obtained with such a taper.

The machining of the center conductor taper is a relatively complicated process and a linear approximation can be used to produce a more economical design approximating the desirable impedance characteristics of the more complicated taper. A cross section of the linear taper is shown in Figure 24 along with the input VSWR of the linearly tapered line section as a function of frequency. Notice that except for the region around 10 Gc, results equivalent to those obtained with the complicated taper are obtained with the

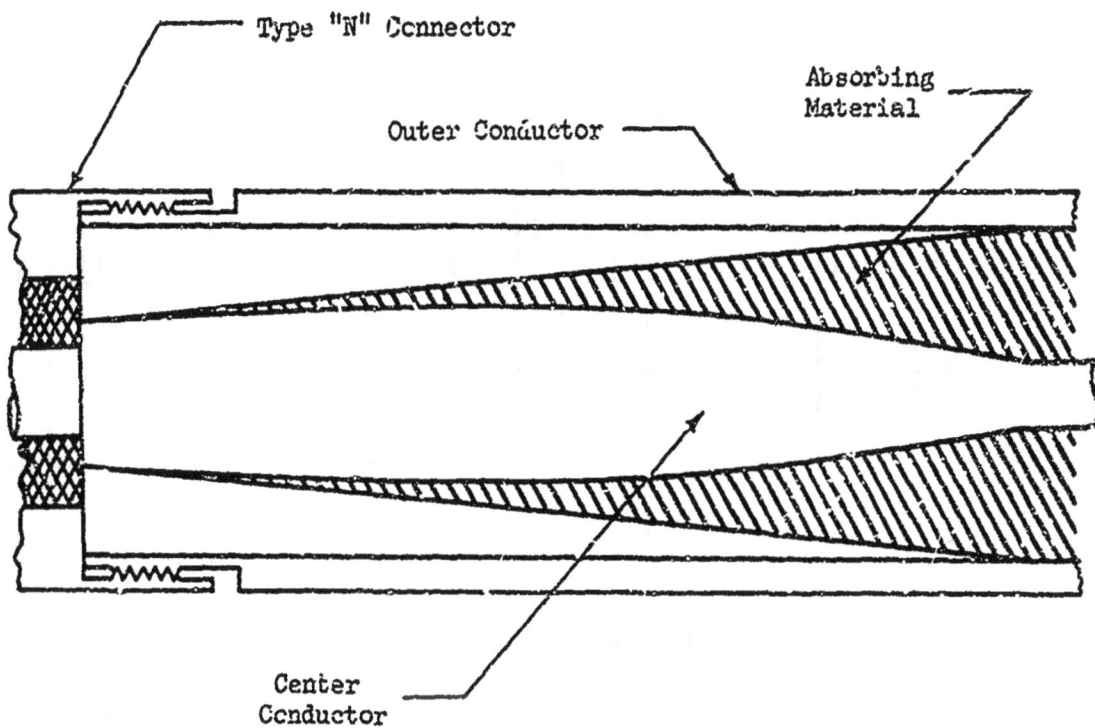

Figure 23. Cross Sectional View of Matching Taper.

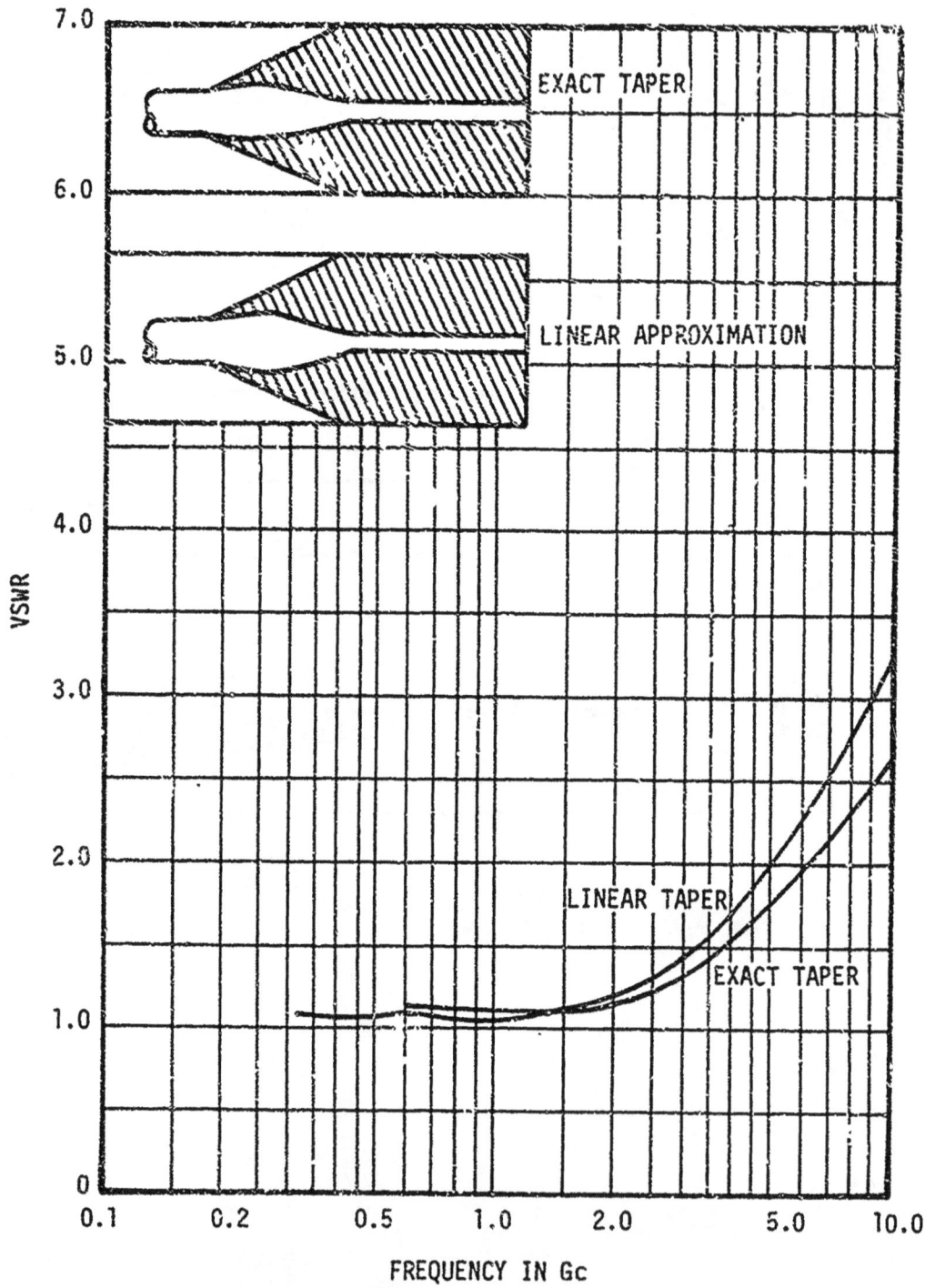

Figure 24. Effect of Linear Approximation on VSWR.

approximating taper, indicating that the linear approximation is adequate for most applications.

If a large number of filters are to be constructed, it would probably be more economical to cast rather than to machine the center conductor in the desired shape and to injection mold the dielectric material in the desired taper in the dielectric space between the center and outer conductors. In this case it would probably be best to construct the molds in the more complicated tapered shape to get the best performance.

2.1.4 Filter Design Procedure

Typical applications of lossy filters might be to suppress a high order harmonic output from a transmitter or to reject out of band signals appearing near a spurious response of a receiver. In either case, the filter is required to produce a certain minimum attenuation at all frequencies higher than some given frequency. For example, suppose a minimum attenuation of 30 db was required at 1 Gc to protect a receiver tuned to 100 Mc from interference generated by a nearby radar transmitter. Referring to Figure 3, it is determined that the standard 7.2 cm length of a 6:1 E grade carbonyl iron material will provide 23 db attenuation at 1 Gc. The length, ℓ, required for 30 db attenuation is then

$$\ell = (30/23)(7.2) = 9.4 \text{ cm} \qquad (16)$$

The passband insertion loss is also given by Figure 3 as approx-imately 0.5 db for 7.2 cm. The reflection loss correction is, from Equation (10),

$$L_R = 40 \log_{10} \left| \sqrt{4.93/19.17} + 1/2 \right|$$

$$- 10 \log_{10} \left[(4)(4.93)/19.17 \right] \approx 0 \quad . \tag{17}$$

The passband loss, L_{PB}, of a 9.4 cm length is therefore equal to

$$L_{PB} = (9.4/7.2)(0.5) = 0.65 \text{ db} \quad . \tag{18}$$

If, instead of carbonyl iron, a mill scale material had been selected, say a 4:1 ratio with No. 4 size particles, the loss of a 7.2 cm section at 1 Gc is given by Figure 10 as 19.5 db. The required length is then

$$\ell = (30/19.5)(7.2) = 11.1 \text{ cm} \quad . \tag{19}$$

The passband reflection loss correction is computed using Equation (10). From Table 5, the values for μ_r' and ϵ_r' at 400 Mc are $\mu_r' = 1.74$ and $\epsilon_r' = 16.8$. Since the values of μ_r' and ϵ_r' are relatively constant below 400 Mc, the 400 Mc values are assumed to hold at 100 Mc. Consequently, the reflection loss correction is

$$L_R = 40 \log_{10}(\sqrt{1.74/16.8} + 1/2)$$

$$-10 \log_{10} \ (4)(1.74)(16.8) \ = 0.4 \text{ db} \quad . \tag{20}$$

From Figure 10, the loss of the standard section at 100 Mc is 4.0 db. Therefore, the actual passband loss of the standard line under matched conditions is

$$L_{PB} = 4.0 - 0.4 = 3.6 \text{ db} \quad , \tag{21}$$

for a 11.1 cm length, the total passband loss is

$$L_{PB(Total)} = (11.1/7.2)(3.6) = 5.6 \text{ db} \quad . \tag{22}$$

The dimensions of the center and outer conductors must be modified from those used in the standard test section if the passband loss of 5.6 db is to be obtained. Using Equation (15) gives

$$\frac{r_1}{r_2} = e^{(5/6)(\sqrt{16.8/1.74})} = e^{(2.59)} = 13.4 \quad . \tag{23}$$

Since the outer conductor of the standard section mates well with type N fittings, the outer conductor should be held fixed and the radius of the inner conductor reduced until the correct ratio of dimensions is obtained. The required inner conductor radius is

54

$$r_2 = 0.28/13.4 = 0.0208 \text{ inches} \qquad (24)$$

The necessary radius is closely approximated by that of a No. 18 gauge bare copper wire. The resulting design of a low-pass lossy line section is sketched in Figure 25.

2.1.5 Cascading Lossy and Reactive Filters

For many low-pass filter applications, attenuation slopes such as those shown in Figure 3 are not sufficiently steep, and a conventional reactive filter must be used in cascade with the lossy line section to supply a steeper attenuation characteristic. The combination of the conventional reactive filter and a lossy section can provide an overall characteristic which has the rapid cutoff slope of the reactive filter as well as the high stop-band attenuation of the lossy section. In addition to the desirable stop-band attenuation performance, the combined filter also has a stop-band impedance which, except for the region near the cutoff frequency of the reactive filter, is set by dissipation in the lossy dielectric material rather by the reflection coefficient of the reactive filter.

Although in most applications the reactive input impedance of a filter in the region near cutoff is no particular disadvantage, it is sometimes necessary that the filter input impedance be resistive over the entire pass and stop-bands. In such a circumstance, compensation for the mismatch at the input terminals of the reactive low-pass filter can be obtained by placing a terminated high-pass

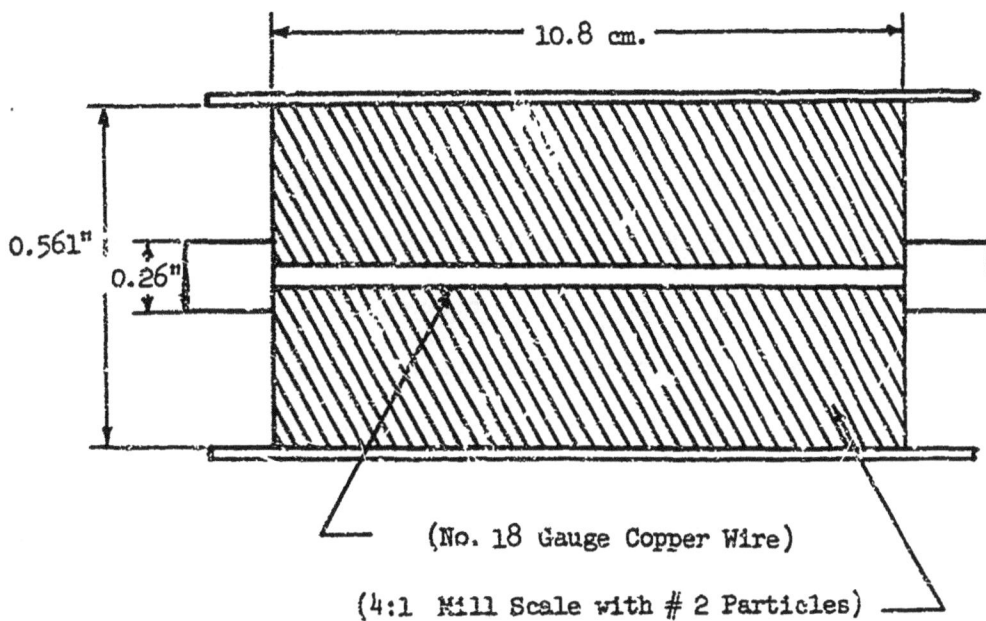

Figure 25. Lossy Line Filter Design Example.

56

filter in parallel with input terminals of the low-pass filter.
It has been shown[2] that the use of a Butterworth attenuation function in the design of both the low-pass and high-pass reactive filters can produce the required compensation of the reactive component of the input impedance of the low-pass filter.

When a lossy line section is used in cascade with a conventional low-pass filter, the passband insertion loss can be minimized by the proper choice of the dielectric material. However, there is always some passband loss introduced by the lossy dielectric. Such passband losses can be reduced by designing the reactive filter to have as wide a region as possible between the low-pass cutoff frequency and the first spurious passband, so that a minimum of lossy material is needed to provide the required stopband attenuation.

An example of the improvement in stop-band attenuation that can be gained by preceding a reactive filter with a lossy line section is illustrated in Figure 26. Figure 26(a) shows the performance of a typical reactive low-pass filter constructed with lumped constant elements. The rapid cutoff at 400 Mc is followed by a high attenuation region between 400 Mc and 3 Gc, but at frequencies above 3 Gc, the attenuation is greatly reduced. If the same low-pass filter is preceded by a 5 cm section of coaxial line whose dielectric space is filled with a 6:1 ratio of iron to epoxy dielectric material, the attenuation characteristic is altered to that shown in Figure 26(b). The addition of the lossy section has increased the passband attenuation only slightly, but the

57

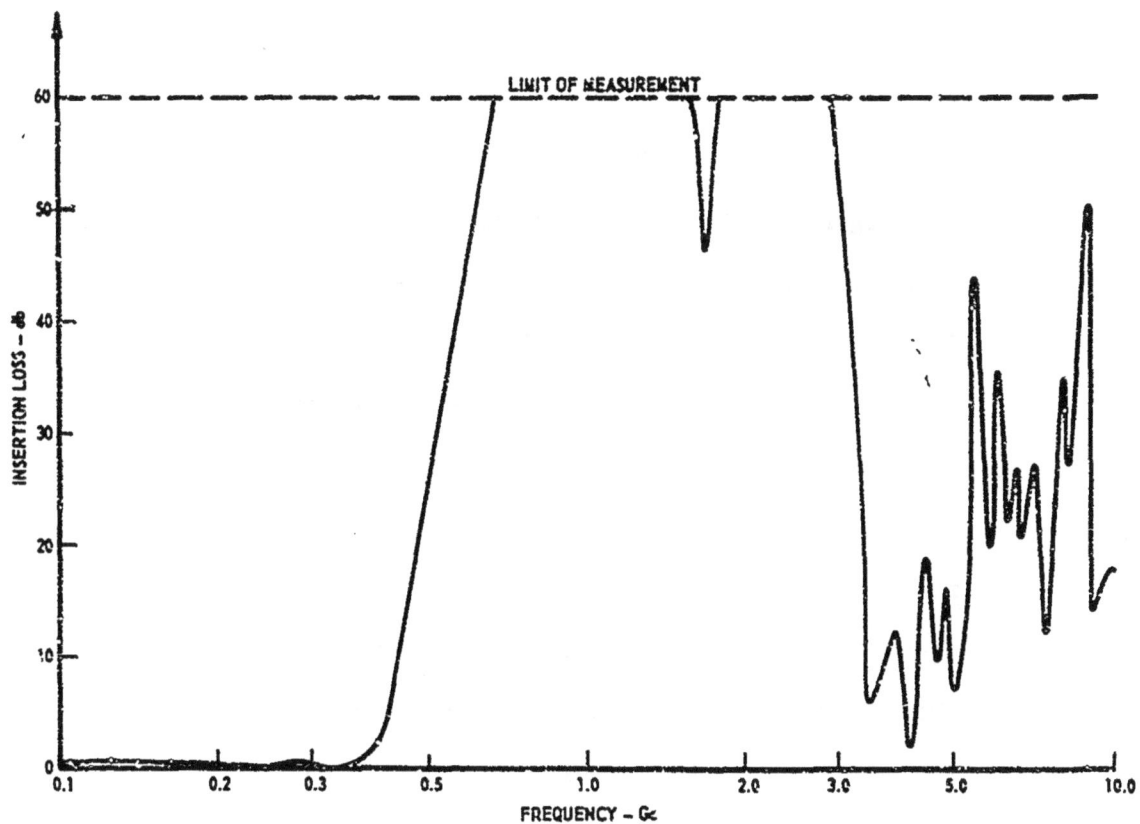

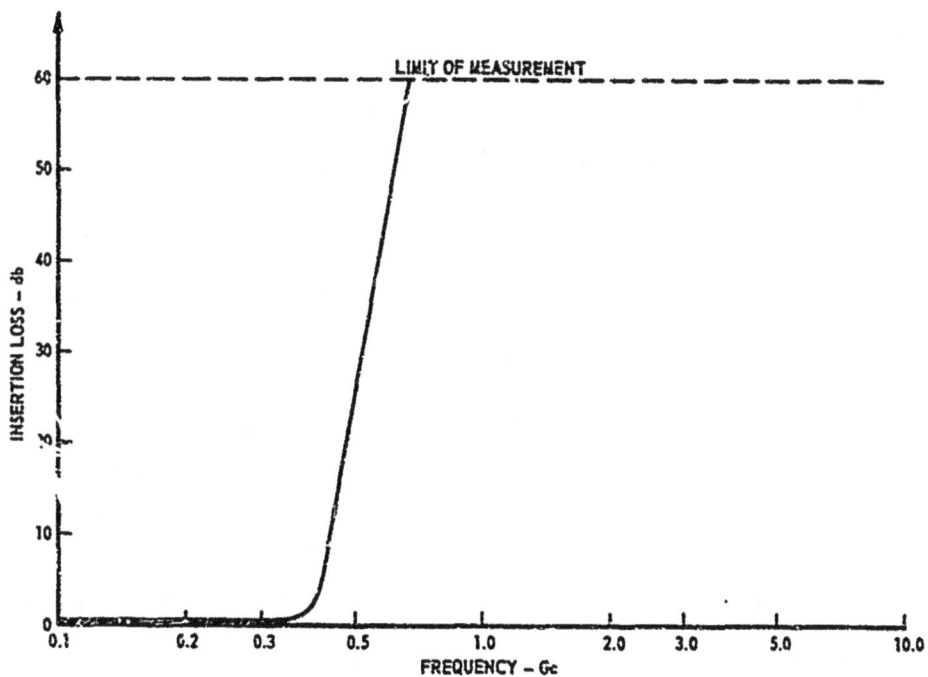

Figure 26. Low-Pass Filter Loss Characteristics.

58

stop-band attenuation has been increased to greater than 60 db.

Figure 27 illustrates the application of a lossy line section to the suppression of spurious responses of a bandpass filter. The frequency response of a typical single cavity coaxial filter is shown in Figure 27(a). Although the main response of the filter exhibits the desired narrow passband, the low attenuation at frequencies above 1 Gc renders the filter almost useless for the rejection of signals at frequencies in this region. When the cavity filter is combined with a low-pass reactive filter and a short section of lossy line, the required high attenuation at frequencies outside the desired passband is obtained, as shown in Figure 27(b). The reactive low-pass filter is used to suppress the large response at 600 Mc, since the attenuation slope of the lossy section is not sufficiently high to give any appreciable attenuation at 600 Mc without producing a significant increase in the insertion loss at the tuned frequency of 200 Mc. However, the reactive low-pass filter alone is not sufficient to suppress the almost continuous passband of the cavity filter at frequencies above 1 Gc, since the low-pass filter itself has regions of low attenuation in the same frequency range. Consequently, the combination of all three filters, i.e., the bandpass filter, the low-pass filter, and the lossy line section, is necessary to produce the desired stop-band performance.

EXCEEDS LIMIT OF MEASUREMENT

REJECTION CHARACTERISTICS
OF UHF CAVITY PRESELECTOR
TUNED TO 200 Mc

ATTENUATION – db

EXCEEDS LIMIT OF MEASUREMENT

REJECTION CHARACTERISTICS OF UHF
PRESELECTOR CONSISTING OF CAVITY, LOW
PASS FILTER, AND LOSSY LINE SECTIONS

ATTENUATION – db

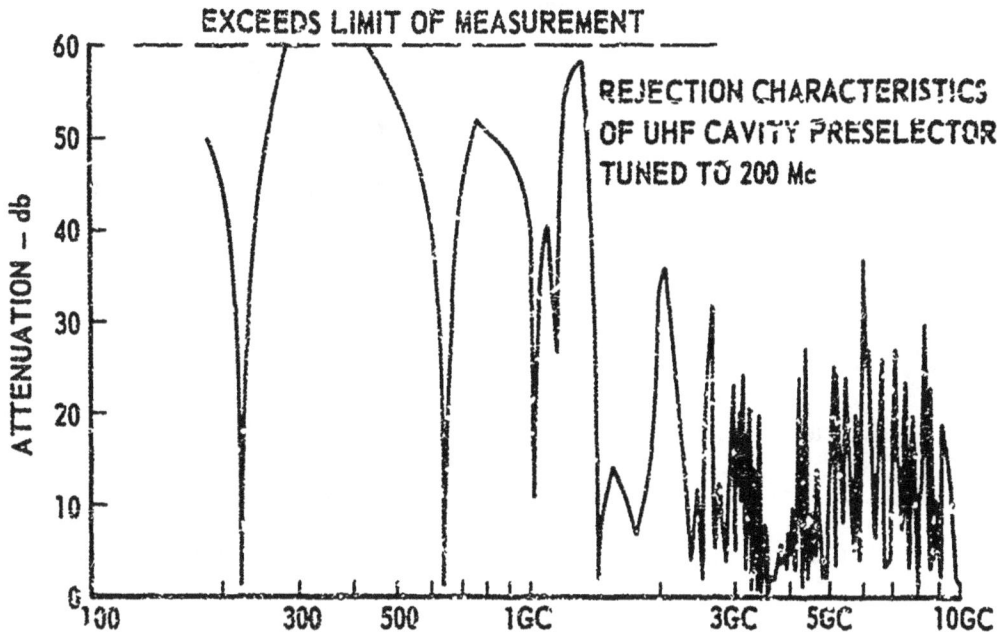

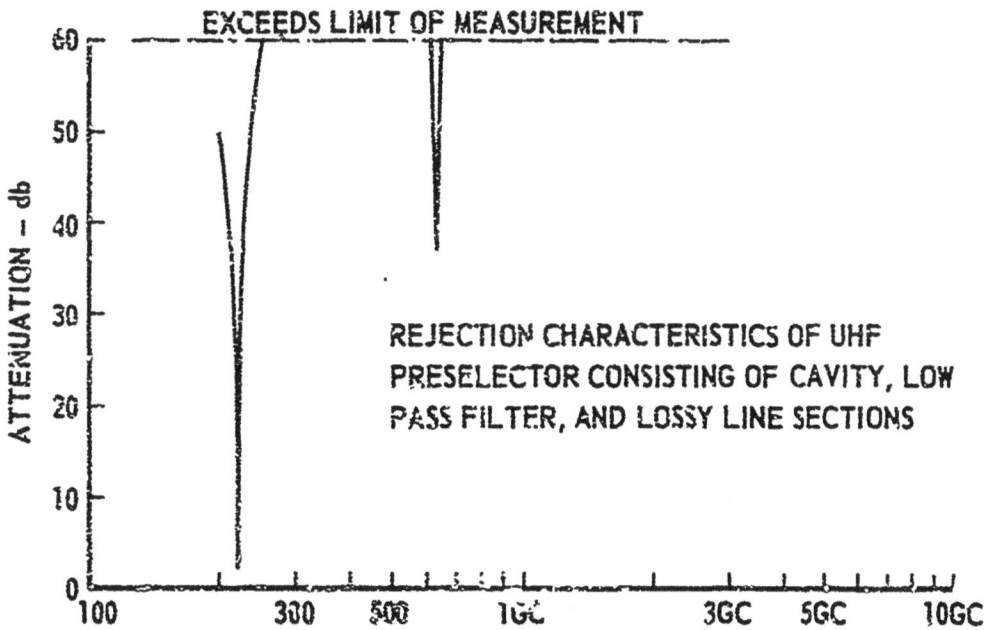

Figure 27. Preselector Loss Characteristics.

2.1.6 Integrated Filters

In some instances, the cascading of a lossy line section
with a conventional reactive filter results in a filter whose
dimensions may be unwieldy. In these situations, a considerable
reduction in filter size can be obtained by filling the dielectric
space in a conventional coaxial filter with a lossy material. This
dimensional reduction is really twofold. First, because the necessity of providing separate lossy and reactive filters is avoided,
and second, because the permeability and dielectric constant of the
dielectric material increases the per unit inductance and capacitance
of the reactive filter sections. Some idea of the size reduction
obtained with this technique is shown in the photograph of Figure
28 which shows a conventional low-pass reactive filter and an
integrated filter of slightly lower cutoff frequency. The cutoff
attenuation slopes of the two filters are approximately the same,
but the shorter filter displays superior stop-band performance.

One disadvantage of combining the lossy and the reactive
sections into one integral filter is that the stop-band input
impedance is no longer set by the dissipation losses in the dielectric material as was the case when the lossy section was separate
from and preceded the reactive filter. Instead, the stop-band
input impedance is essentially reactive since an input signal
encounters a reactive element of the filter before any appreciable
loss takes place. If both small size and matched stop-band input
impedances are important, a compromise is possible in which a short

61

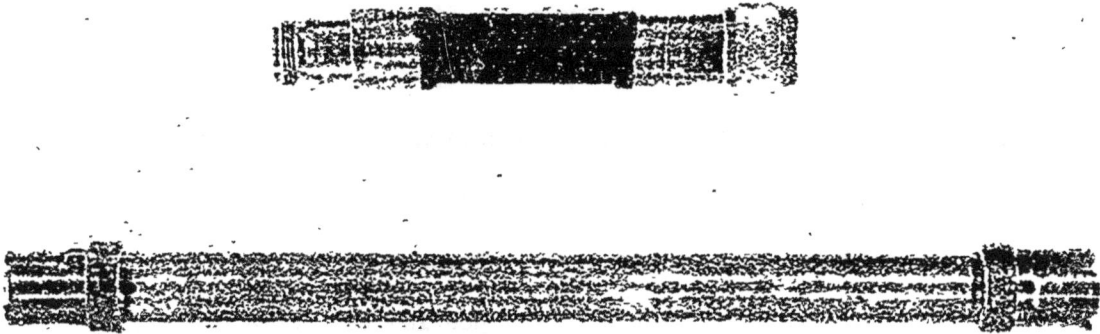

Figure 28. Reactive and Lossy Filters.

tapered section of lossy material is placed around the input center
conductor of the filter to provide an adequate stop-band match,
with additional material being placed around the reactive elements
of the filter to provide high stop-band attenuation. If the
overall passband attenuation of the filter is to be kept as small
as when no lossy material is placed on the input center conductor,
a reduction in the ratio of iron to epoxy is necessary in order to
provide a smaller passband attenuation per unit length due to the
lossy material.

2.1.7 Balanced Transmission Line Applications

The use of the carbonyl iron-epoxy material to provide a
frequency sensitive attenuation characteristic is readily applicable
to balanced as well as to coaxial transmission lines. Several 300
ohm transmission line sections were prepared by placing a section
of 300 ohm "twin lead" in a mold and pouring the lossy mixture
around it. A typical insertion loss curve for a "twin lead" transmission line with 3 inch long and 1/8 inch thick coating of 6:1
carbonyl iron lossy material applied is shown in Figure 29(b). A
considerable increase in attenuation per unit length is realized
if the insulation is removed from the section of "twin lead" that
is to be coated with the lossy material. The effect of removing
the "twin lead" insulation is shown by curve (a) of Figure 29. If
a steeper cutoff attenuation slope than that provided by the lossy
section alone is desired, a reactive filter filled with lossy

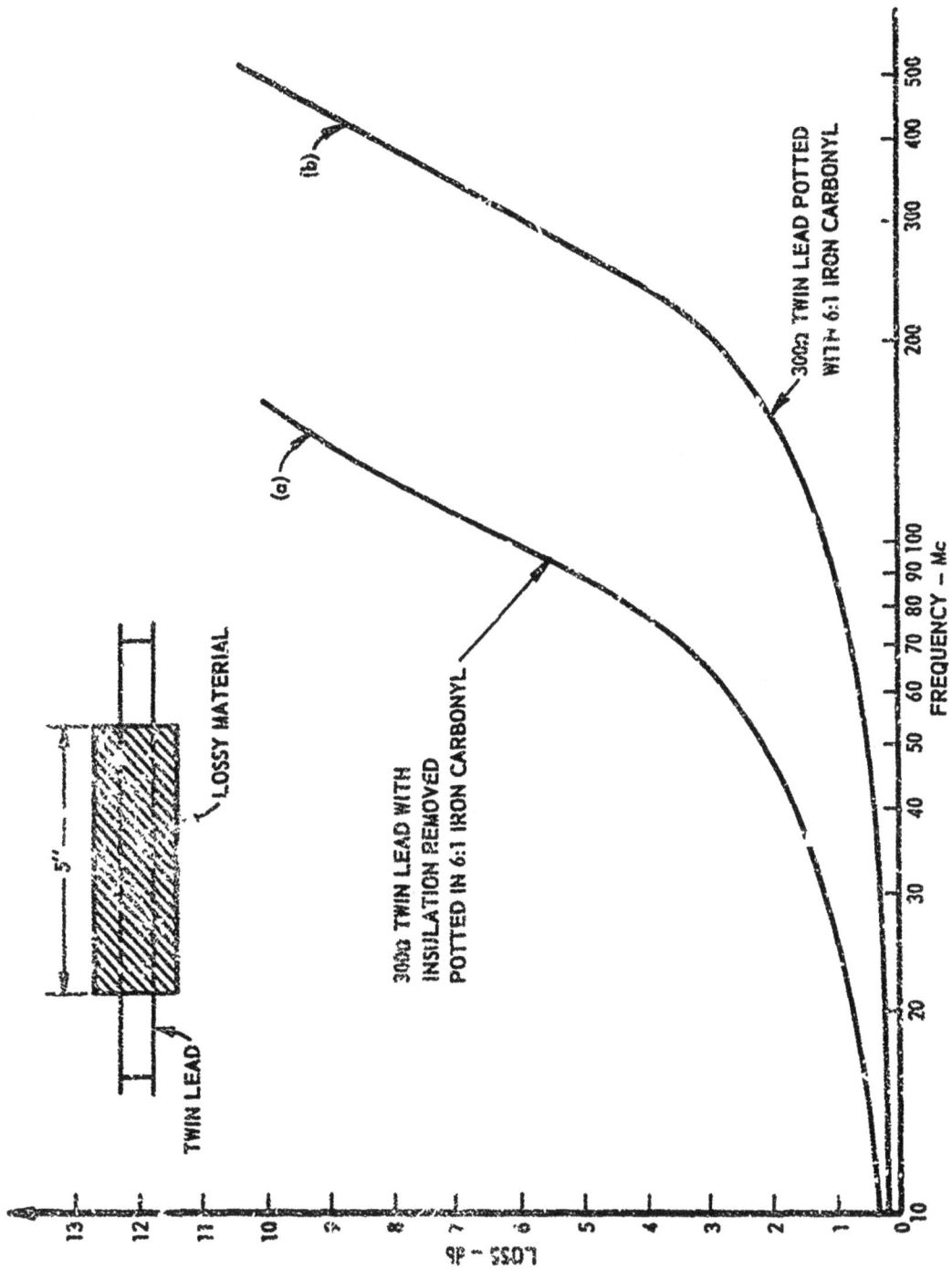

Figure 29. Balanced Transmission Line Characteristics.

material may be used in a manner similar to that already described for coaxial systems. A low-pass filter of this type was designed to be used as a filter for VHF television sets. The attenuation is low throughout the VHF-television band and rises steeply at frequencies above 250 Mc. Such a filter can serve to reduce interference to television sets by high power sources operating near one of the UHF spurious responses of a commercial television set. The insertion loss characteristic of the balanced transmission line filter constructed is shown in Figure 30. The necessary inductance for the filter was supplied by the two conductors of the "twin lead" with the insulation removed. Small mica capacitors were placed in parallel along the "twin lead" to provide the required capacitance. The assembly was then placed in a thin plastic box which served as a mold and the box was filled with an iron and epoxy mixture. This type of filter is attractive in that it is simple to construct and uses very few components.

2.1.8 Lossy Filters for HF

Low-pass filters with cutoff frequencies in the HF range often have spurious passbands in the low UHF or high VHF region. If a carbonyl iron lossy line section is used to provide significant attenuation in these spurious passband regions, the length of line required is excessive. A more practical line length is obtained if another dielectric material is used to supply the attenuation. For example, a T-1 ferrite with a 4:1 ratio has a 20 db loss at 400 Mc

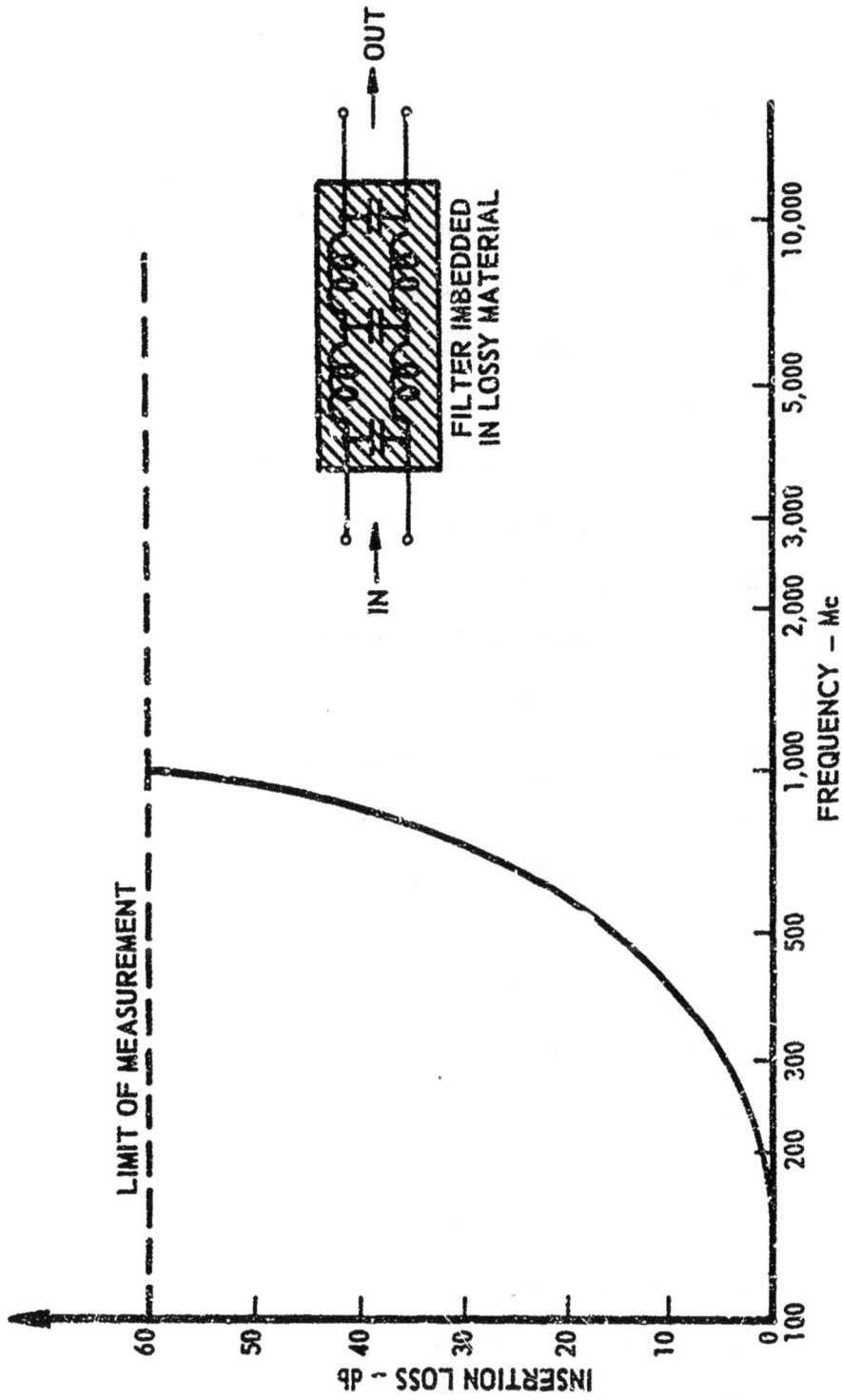

Figure 30. Balanced Integrated Filter.

in a length of 7.2 cm. A decrease in the physical length of the lossy line section required to obtain a given attenuation can also be obtained by coiling the center conductor of the line section into a helix. The overall loss is then given by the attenuation per unit length times the total length of wire used in forming the helix. Another material which can produce high loss at VHF and UHF frequencies is the Glidden Company magnetite M-180. A 6:1 ratio of M-180 to epoxy produces 15 db attenuation in a standard length at 400 Mc. If the construction of a large number of lossy line sections is contemplated, the use of M-180 to replace the powdered ferrite should be considered from a cost standpoint. At a cost of 9 cents per pound for the magnetite as compared to $10.00 per pound of the ferrite powder, it is possible that a significant per unit cost reduction can be obtained by using the M180 material.

The technique of applying the lossy line sections to the improvement of HF filters is essentially the same as that used in conjunction with UHF filters. The most direct approach is the use of a lossy line section in cascade with a reactive filter. Proper selection of the lossy material then provides the desired uniformly high stop-band attenuation without appreciably increasing the pass-band loss. Integrated designs are also feasible. If the reactive low-pass filter is constructed from lumped circuit elements in a container, the container can be filled with the lossy material to suppress undesirable UHF and VHF passbands. When such a procedure is used, element values must be adjusted to compensate for the change in the values of the components that takes place when they are surrounded by the dielectric material. A variation of this technique has been used with success by this project as well as by others.[3] Instead of filling the entire container with dielectric material, only the inductances are surrounded with lossy material. Since inductors are generally the worst offenders in producing spurious responses in a lumped constant filter, the use of the lossy dielectric material around the inductors greatly reduces these spurious effects. Care must be taken to use only low inductance capacitors and careful shielding between sections if this simplified suppression technique is to be of maximum effectiveness. Reference 3 gives several examples of successful applications of the technique using commercially available inductances.

2.1.9 High Power Lossy Filters

The effect of temperature on the loss characteristics of a carbonyl iron dielectric material was examined by cycling the temperature of a lossy line section over the temperature range of -50° C to +100° C. Over this range no significant changes in the loss characteristics were observed.

The ability of the lossy material to operate at elevated temperatures without appreciable change in its attenuation characteristics suggests its use in high power filter applications. When the entire dielectric space of the transmission line is filled

68

with lossy material, peak power limitations are set primarily by the breakdown voltage of the material while average power limitations are set by the allowable temperature rise in the material. Since the power dissipated in a homogeneous lossy material varies exponentially with respect to distance from the input terminal, the amount of lossy material per unit length of transmission line must vary with length if the power dissipated in each unit length of line is to be equal. A convenient means of equalizing the power dissipation is to use a short taper at the input to the lossy section which provides a gradually increasing loss per unit length. When this power equalization has been accomplished, a sufficiently long line must be used to provide adequate surface area to prevent an excessive rise in temperature. The taper also has the additional advantage of providing a smooth transition from the air dielectric to the lossy dielectric with a consequent reduction in the input VSWR.

To establish the peak power limitations of typical lossy dielectric materials, an examination was made of those factors which influence the breakdown potential of the lossy material. Since the breakdown potential of the epoxy binder is considerably higher than that of the powdered iron alone, it is reasonable to expect that the breakdown potential decreases as the percentage of iron in the material increases. The curve of Figure 31(a) verifies this expectation. On the basis of Figure 31(a) it should be possible to provide any breakdown voltage less than the breakdown rating

(a)

BREAKDOWN VOLTAGE in KV.

MIX RATIO

(b)

Silver
Electrodes

Test Sample

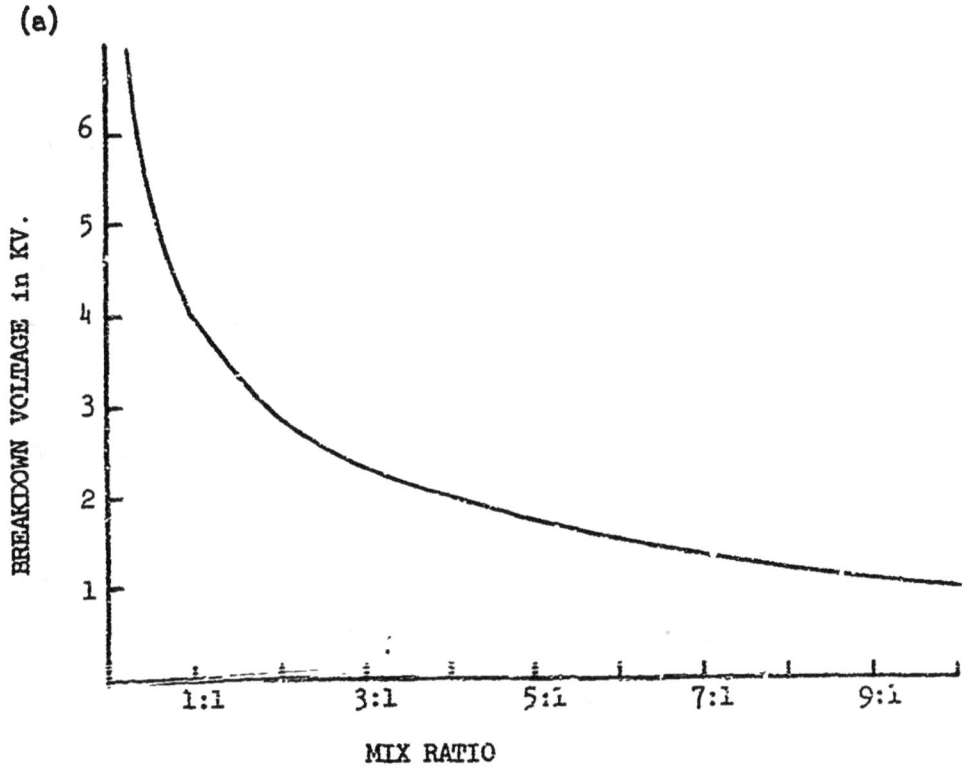

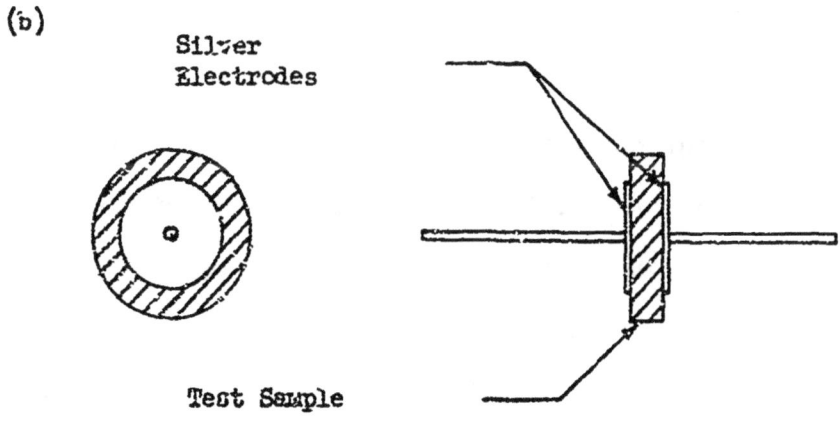

Figure 31. Voltage Breakdown Versus Mix Ratio.

70

of the epoxy binder by using a sufficiently low mix ratio, with the required overall attenuation being obtained by an increased length of the lossy section to compensate for the lower loss per unit length. However, there is a lower limit to the mix ratio that should be used. This limit is set by the losses of the epoxy binder which tend to swamp out those of the powdered iron when the mix ratio is much below 2:1. The dielectric loss in the epoxy is not nearly so frequency sensitive as that of the iron so that the cutoff slope of the frequency versus attenuation characteristic is appreciably reduced at very low mix ratios. If it becomes necessary to use mix ratios less than 2:1, a lower loss epoxy should be used. Such epoxies are available but they are more expensive and generally require a more complicated curing schedule than the Shell Epon 828 that was used in the construction of all the transmission line filters tested.

Although Figure 31(a) shows the general variation of breakdown potential with mix ratio, a considerable variation of breakdown potential between test samples of the same mix ratio was noted. Additional tests were made in an attempt to identify the factors in the technique of sample preparation which might influence the dielectric breakdown potential. The two primary sources of variation of the breakdown potential were thought to be the lack of uniform mixing of the iron and epoxy and the presence of small air bubbles introduced in the mixing process. Test samples were prepared by filling cylindrical molds with the lossy dielectric material. The samples were mechanically vibrated for varying lengths of time as the epoxy was setting up. This mechanical vibration produced more complete mixing and caused the trapped air bubbles to rise to the top of the mixture. When the top layer was removed, the remaining sample was relatively free of air bubbles. Other samples were allowed to harden under a vacuum to remove the trapped bubbles. Small cylindrical test specimens were machined from the test samples and connecting leads were attached with conducting cement. The sketch of Figure 31(b) shows the dimensions of a typical test specimen. A variable DC voltage was applied across the sample and the voltage increased until breakdown occurred. The results of tests on 62 samples are summarized in Table 7.

TABLE 7

RESULTS OF BREAKDOWN TESTING

Material	Lowest Breakdown Voltage (kv/cm)	Comments
6:1E - Carbonyl Iron	3.5	Not vibrated
6:1E - Carbonyl Iron	2.9	Vibrated 1/2 hour
6:1E - Carbonyl Iron	1.78	Vibrated 1 hour
6:1E - Carbonyl Iron	1.95	(Under vac while setting)
6:1E - Carbonyl Iron	0.65	(Heated to 100° C while setting up)
4:1E - Carbonyl Iron	3.8	Vibrated 1 hour
2:1E - Carbonyl Iron	1.9	Vibrated 1 hour
6:1SF - Carbonyl Iron	3.8	Not vibrated
6:1TH - Carbonyl Iron	3.5	Not vibrated
4:1 - E Ferrite	>12	Not vibrated

Although there was still some variation of the breakdown potential between samples prepared in the same manner, the results indicate that vibration of the samples lowers the breakdown voltage. Nevertheless, some vibration is necessary to remove air bubbles. A compromise between these factors calls for thorough stirring of the iron and epoxy, and subsequent vibration of the sample for one-half hour to reliably provide a breakdown exceeding 2 kv per inch. The coaxial lines normally used to construct lossy sections have a dielectric space approximately one-fourth inch wide. Consequently, the breakdown potential of these lines should exceed 500 volts. On a matched 50 ohm system, 500 volts peak corresponds to an RF power level of 2.5 kw.

It is also possible to increase the breakdown rating of the coaxial line section by placing an air gap between the center conductor and the lossy dielectric material.

The use of such an air gap to establish the required voltage breakdown rating alters both the characteristic impedance and the attenuation constant of the lossy section. In general, the effect of the gap is to reduce both the per unit capacitance and inductance. Since the characteristic impedance is proportional to $(L/C)^{1/2}$, the effect of a small air gap is to raise the characteristic impedance over that of a line with no gap. The attenuation constant is proportional to $(LC)^{1/2}$, and the effect of the gap is to reduce the attenuation constant. The specific variation of the characteristic impedance and attenuation constant can be calculated from the per unit values of L and C. The values

73

of L and C for a coaxial line with a lossy dielectric and the geometry shown in Figure 32 are

$$L = \frac{\mu_o \ell}{2\pi} \left\{ \ln\left(\frac{r_2}{r_1}\right) + \mu_r \ln\left(\frac{r_3}{r_2}\right) \right\} \quad , \tag{25}$$

and

$$C = 2\pi\epsilon_o \ell \left\{ \frac{\epsilon_r}{\epsilon_r \ln\left(\frac{r_2}{r_1}\right) + \ln\left(\frac{r_3}{r_2}\right)} \right\} \quad . \tag{26}$$

For a lossy dielectric, both ϵ_r and μ_r are complex quantities. Under the assumption that only the TEM mode of propagation exists, the loss per unit length is

$$\alpha = \text{Re} \left[\left\{ \mu_o \epsilon_o \right\}^{1/2} \left\{ \ln\left(\frac{r_2}{r_1}\right) + \mu_r \ln\left(\frac{r_3}{r_2}\right) \right\}^{1/2} \right.$$
$$\left. \left\{ \ln\left(\frac{r_2}{r_1}\right) + \left(\frac{1}{\epsilon_r}\right) \ln\left(\frac{r_3}{r_2}\right) \right\}^{-1/2} \right] \quad , \tag{27}$$

and the characteristic impedance is

$$Z_o = \frac{1}{2\pi} \left[\frac{\mu_o}{\epsilon_o} \left\{ \ln\left(\frac{r_2}{r_1}\right) + \mu_r \ln\left(\frac{r_3}{r_2}\right) \right\} \cdot \right.$$
$$\left. \left\{ \ln\left(\frac{r_2}{r_1}\right) + \left(\frac{1}{\epsilon_r}\right) \ln\left(\frac{r_3}{r_2}\right) \right\} \right]^{1/2} \quad . \tag{28}$$

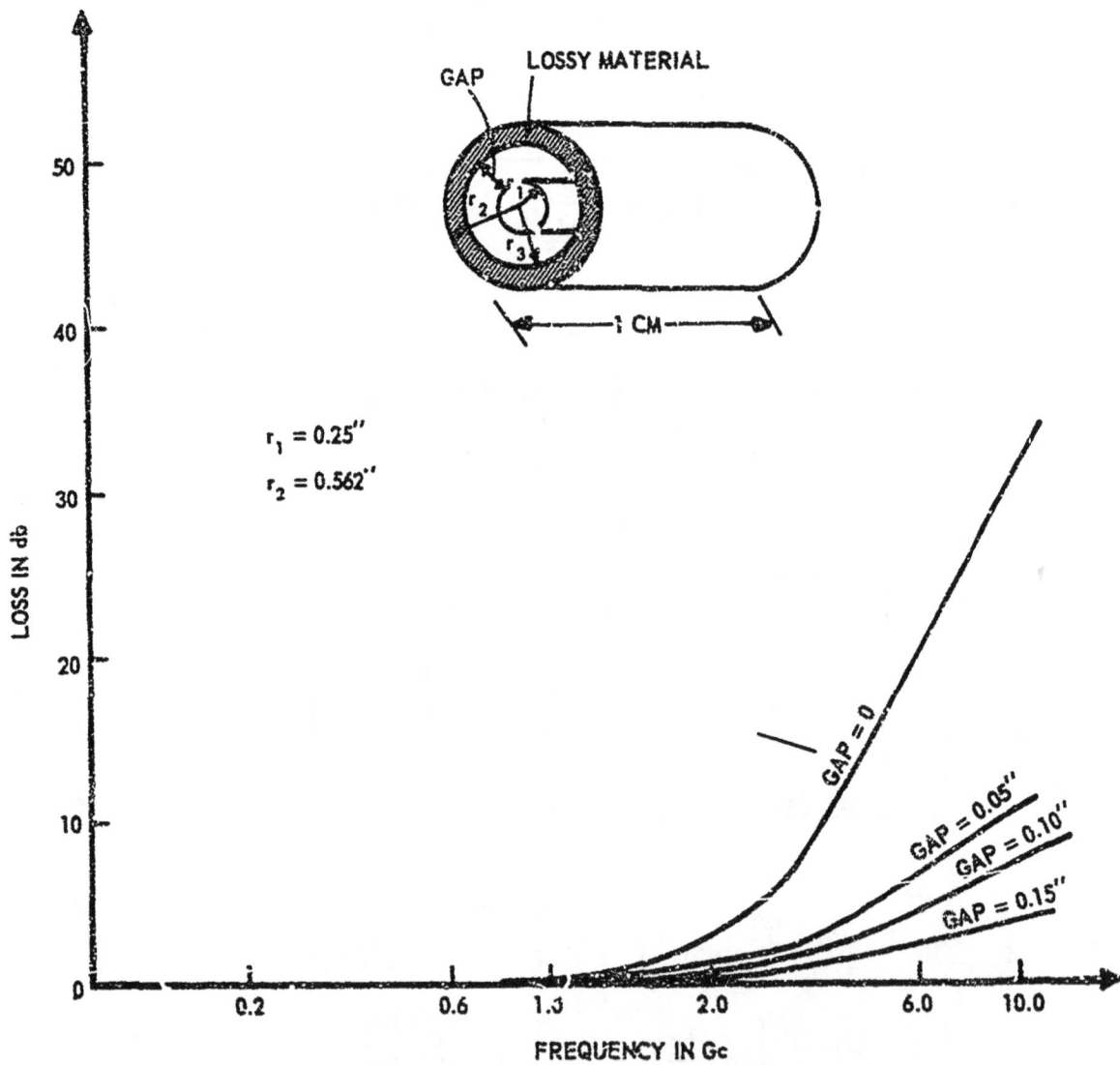

Figure 32. Calculated Effect of Gap on Loss.

Although Equations (27) and (28) can be arranged to express the values of Z_0 and α in terms of μ_r', ϵ_r', δ_ϵ, and δ_μ, the resulting expressions are very cumbersome, and it is simpler to calculate the values of Z_0 and α from Equations (27) and (28) directly.

A typical set of calculated loss versus frequency curves are shown in Figure 32. Equation (27) was evaluated at several frequencies using the dielectric constant and permeability data of Table 1. These calculations were repeated for the several values of gap dimension.

Curves such as those of Figure 32 are readily applicable to the design of low-pass sections using a standard outer conductor with an inner diameter of 9/16-inch. From the operating power level, the necessary air gap is determined. Figure 32 then gives the attenuation in db/cm versus frequency for that particular air gap. If the maximum passband loss is specified, then the length of the section in cm is obtained by dividing the maximum permissible loss by the attenuation/cm as read from Figure 32 at the highest passband frequency. If this length is too short to adequately dissipate the power, then a longer section with a wider gap may be used, or a lower loss material can be selected.

A test section was constructed using a carbonyl iron dielectric material. A small air gap was placed between the center conductor and the dielectric material and the attenuation versus frequency characteristics were measured. The results are shown in Figure 33. Although the general shape of the attenuation curve is similar to

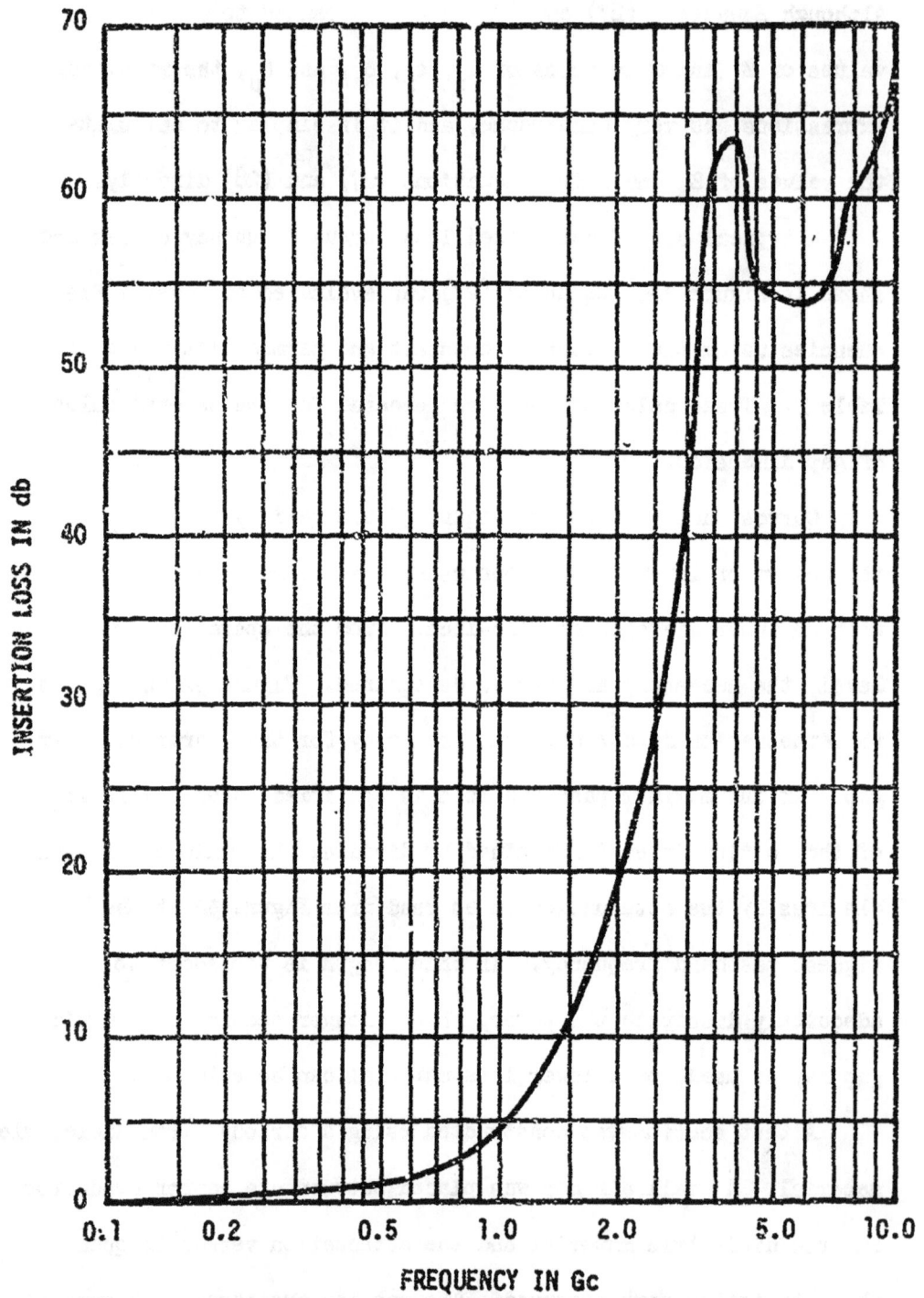

Figure 33. Attenuation of Lossy Filter.

that of the transmission lines containing no gap, there is an undesirable dip in attenuation between 4 Gc and 8 Gc, which had not been observed on lines with no air gap. In the particular line section used, the minimum value of attenuation at the dip is greater than 50 db and would present no serious spurious response problems in most applications.

Because of the unexpected dip that appears in the attenuation curve of Figure 33, an additional investigation was made of the use of an air gap to establish a high breakdown voltage. The results of some insertion loss measurements made on a section of lossy line in which the gap dimension was varied, are shown in the curves of Figure 34. The curve for a completely filled line is included and is in conformance with the previously measured loss characteristics of such a line. The effect of increasingly larger gaps is shown in the other three curves. Notice that the lack of stop-band uniformity becomes increasingly significant as the gap is made larger. If these curves are compared with the theoretical loss curves of Figure 32, it can be seen that this non-uniformity is not predicted by the assumption of TEM propagation. Although it is true that some portion of the variation in stop-band losses may be attributed to the change in the characteristic impedances of the line sections as the gap is varied, such impedance variations are not sufficiently large to account for the difference between the predicted and measured loss curves.

The change in VSWR produced by gapping the line is shown in

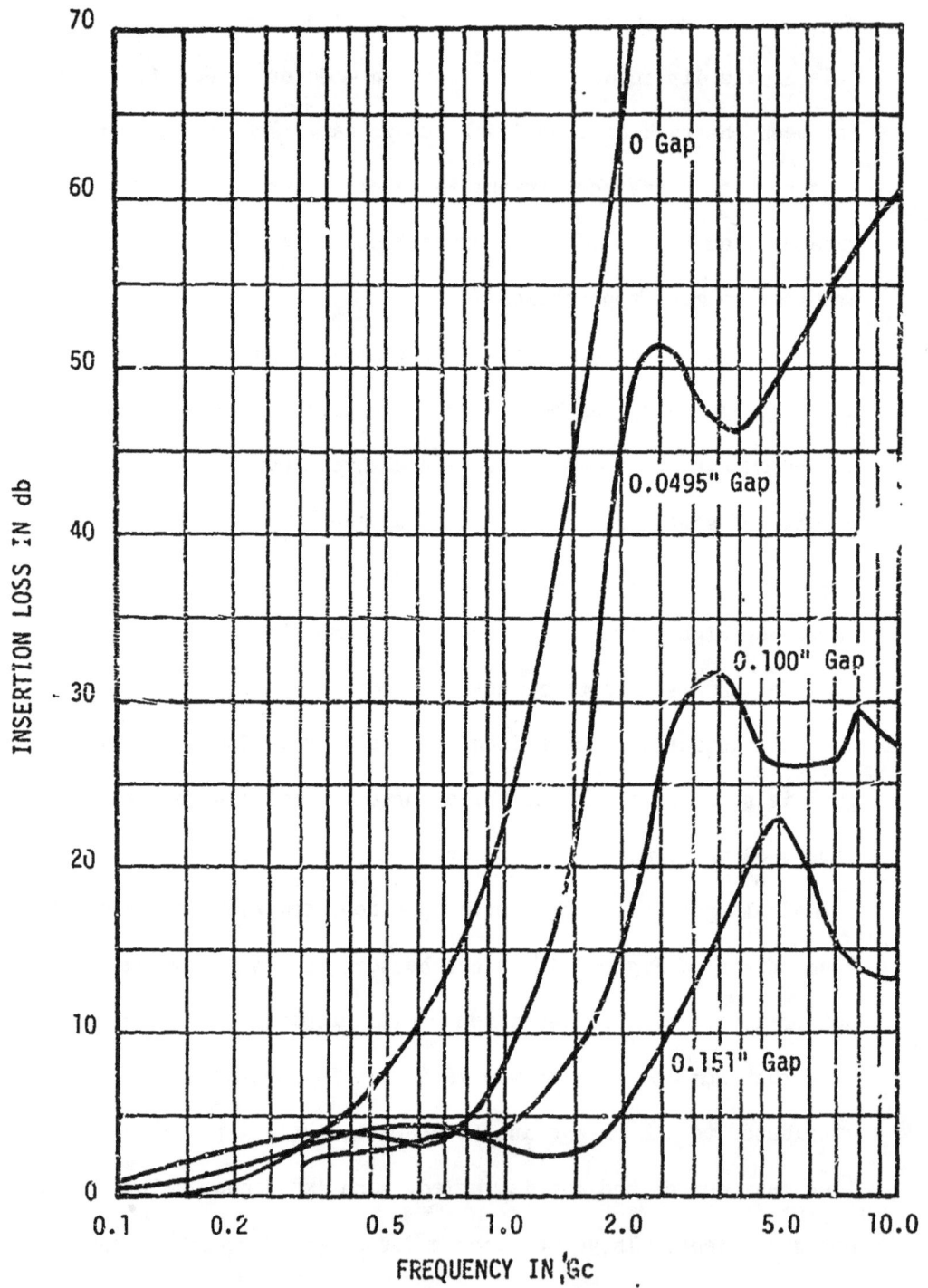

Figure 34. Measured Effect of Gap on Loss.

Figure 35: Notice that the low frequency VSWR for the gapped sections exceeds 2:1, which accounts in part for the high passband insertion loss indicated in Figure 34 for the gapped sections. The zero gap curve shows a steadily rising mismatch as the frequency is increased because no taper was used at the input to the filter.

The results of the loss measurements indicate that the gapping technique is of maximum usefulness for small gaps, and that use of a completely filled dielectric space is the more desireable technique of filter construction. For high power use, the dimensions of the completely filled line sections must be increased sufficiently to provide the necessary breakdown voltage rating in the dielectric material.

Two examples of lossy filters designed to operate at a nominal power level of 100 watts cw are shown in the photograph of Figure 36. These lossy line sections have heat radiating fins to maximize the heat dissipating ability of the filters. Both sections are intended for use in the range 200 to 400 Mc, with the loss through the short section being approximately 1.5 db at 400 Mc while the longer filter has about 2.0 db loss at the same frequency. With a 400 Mc input power level of 100 watts, power lost in the lossy lines is 29.3 watts in the short section of 36.9 watts in the longer section. A more detailed picture of the construction of the lossy sections is given by the sketch of Figure 37 which shows a sectional view of the filter. Notice that the air gap has been placed between the center conductor and the lossy material. This arrangement places the

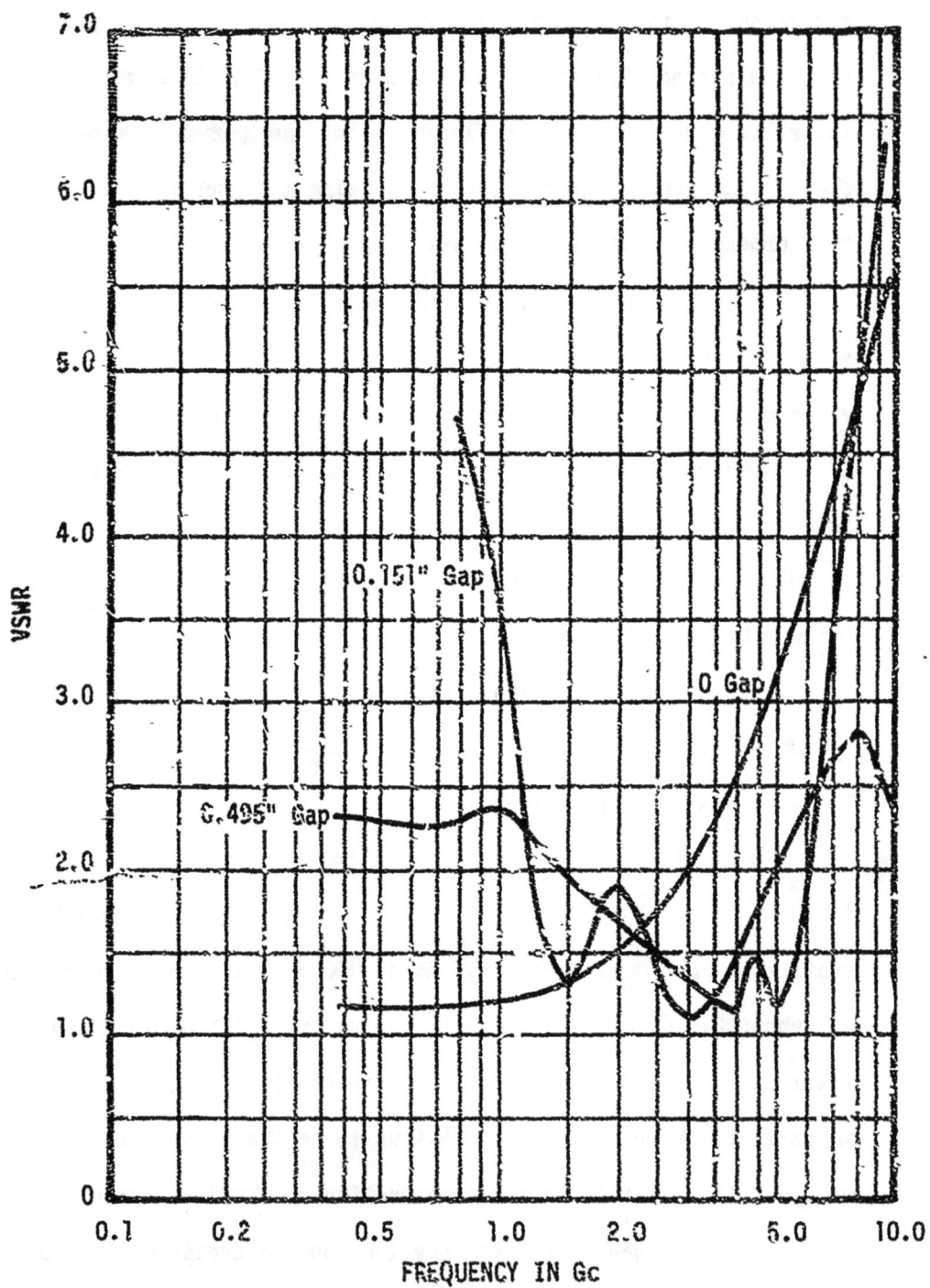

Figure 35. Effect of Gap on VSWR.

81

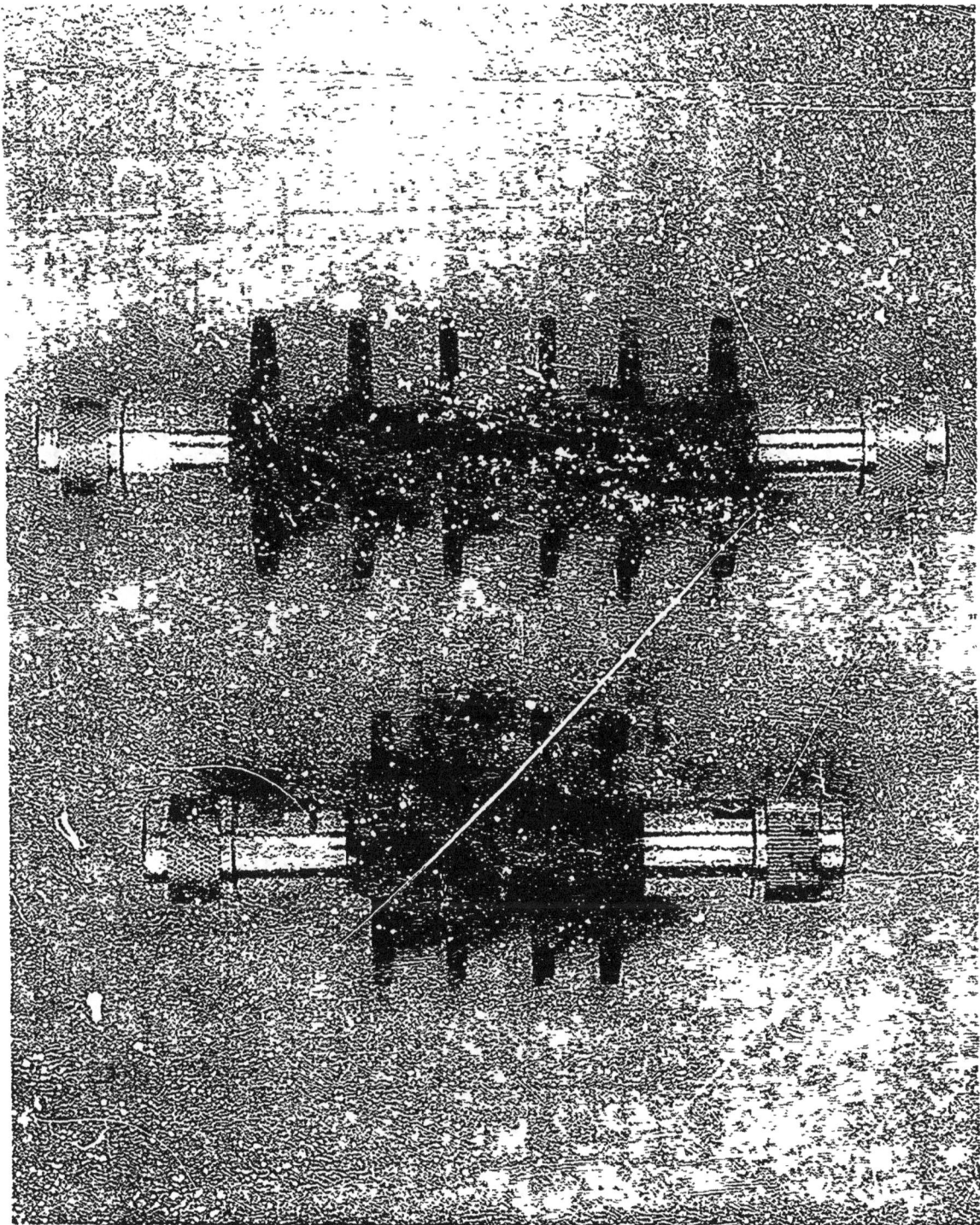

Figure 36. Lossy Line Sections.

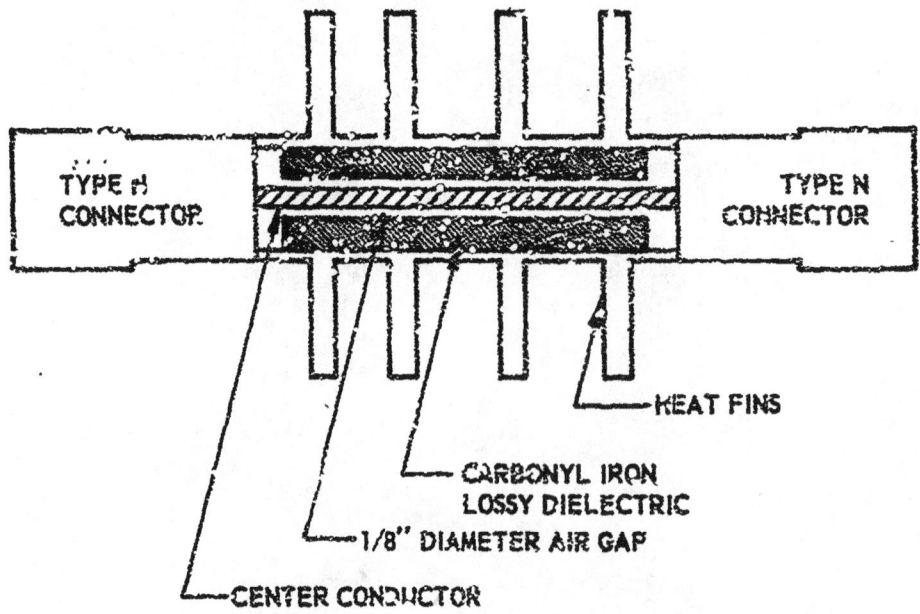

TYPE N CONNECTOR

TYPE N CONNECTOR

HEAT FINS

CARBONYL IRON LOSSY DIELECTRIC

1/8" DIAMETER AIR GAP

CENTER CONDUCTOR

Figure 37.   Lossy Line Construction.

lossy material in close contact with the brass outer conductor and
gives maximum heat transfer from the dielectric material to the
heat dissipating fins. Construction is also simplified by placing
the air gap next to the center conductor since the outer conductor
shell can be poured full of the iron and epoxy mixture which, after
the epoxy has hardened, can be drilled out to provide a clearance
hole for the center conductor and the necessary air gap. Since
the power dissipation requirements for these particular lossy line
sections were relatively low, no attempt was made to taper the air
gap to provide equal power dissipation per unit length, thus
simplifying the construction considerably. Figure 38 illustrates
the manner in which the short lossy section was cascaded with a
conventional short-line coaxial filter to produce a low-pass filter
with both the steep cutoff characteristics of the reactive filter
and the uniformly high stop-band attenuation of the lossy section.
The dotted curve shows the frequency response of the reactive
filter which has serious holes in its stop-band attenuation characteristics. The solid curve illustrates the composite attenuation
curve measured for the combination of the reactive filter and the
lossy section.

Tests of the filter were made using a high power UHF signal
source located at RADC. These tests verified that the low power
(0 dbm) attenuation characteristics were maintained at power levels
of at least 100 watt. Tests were also made to determine if any
nonlinear effects were taking place inside the lossy material which

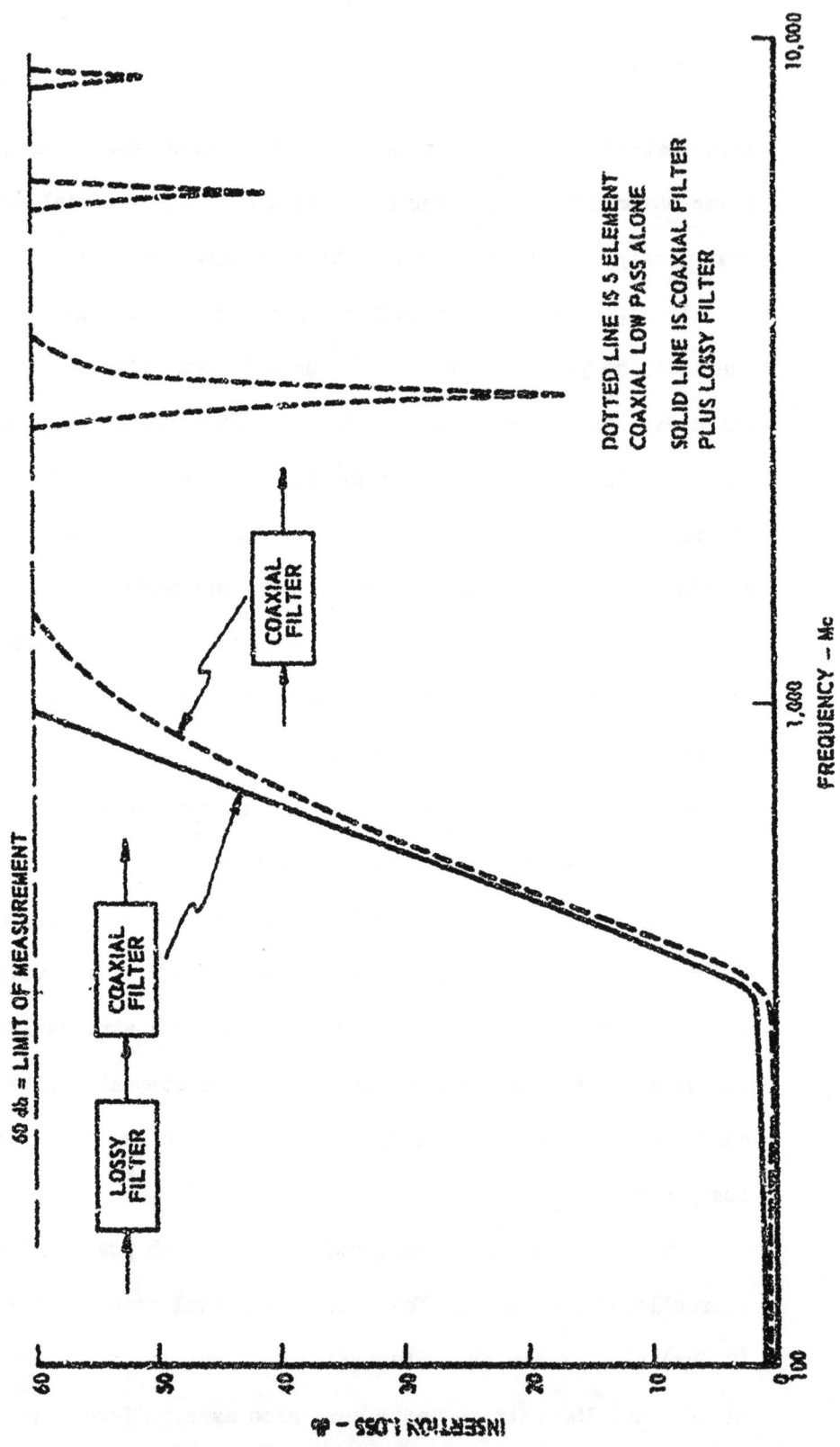

Figure 38. Lossy Line and Reactive Filter Attenuation Functions.

might cause harmonics to be generated. Although the sensitivity of the measuring equipment was such that harmonic levels as low as 100 db below the level of the fundamental could be detected, no harmonic power was observed. Typical results of the use of the combination filter as a harmonic filter for the high power source are illustrated by the curve of Figure 39. The operating conditions and measuring equipment arrangements are shown on the figure.

A lossy filter for use at high power levels was constructed with the dielectric space completely filled with the lossy material. The section was tested at high power levels to verify the fact that there is no drastic difference between the DC breakdown potential and the RF breakdown potential of the dielectric material. The tests were performed using the high power UHF signal source at RADC. The lossy section was operated at UHF power levels up to 400 watts over a period of several hours. No breakdown problems were encountered. The test section was constructed with a 2:1 mix ratio dielectric material to distribute the insertion loss over a longer length. The overall length of the test section was 5 inches. Temperature measurements made on the heat dissipating fins of the lossy section showed an equilibrium temperature of approximately 95° C at an input RF level of 400 watts. Temperatures inside the dielectric were necessarily higher. If operation at higher power levels is required, a longer section with a lower mix ratio will be necessary to prevent excessive temperature rise in the dielectric material. Since, for reasons already stated, mix ratios

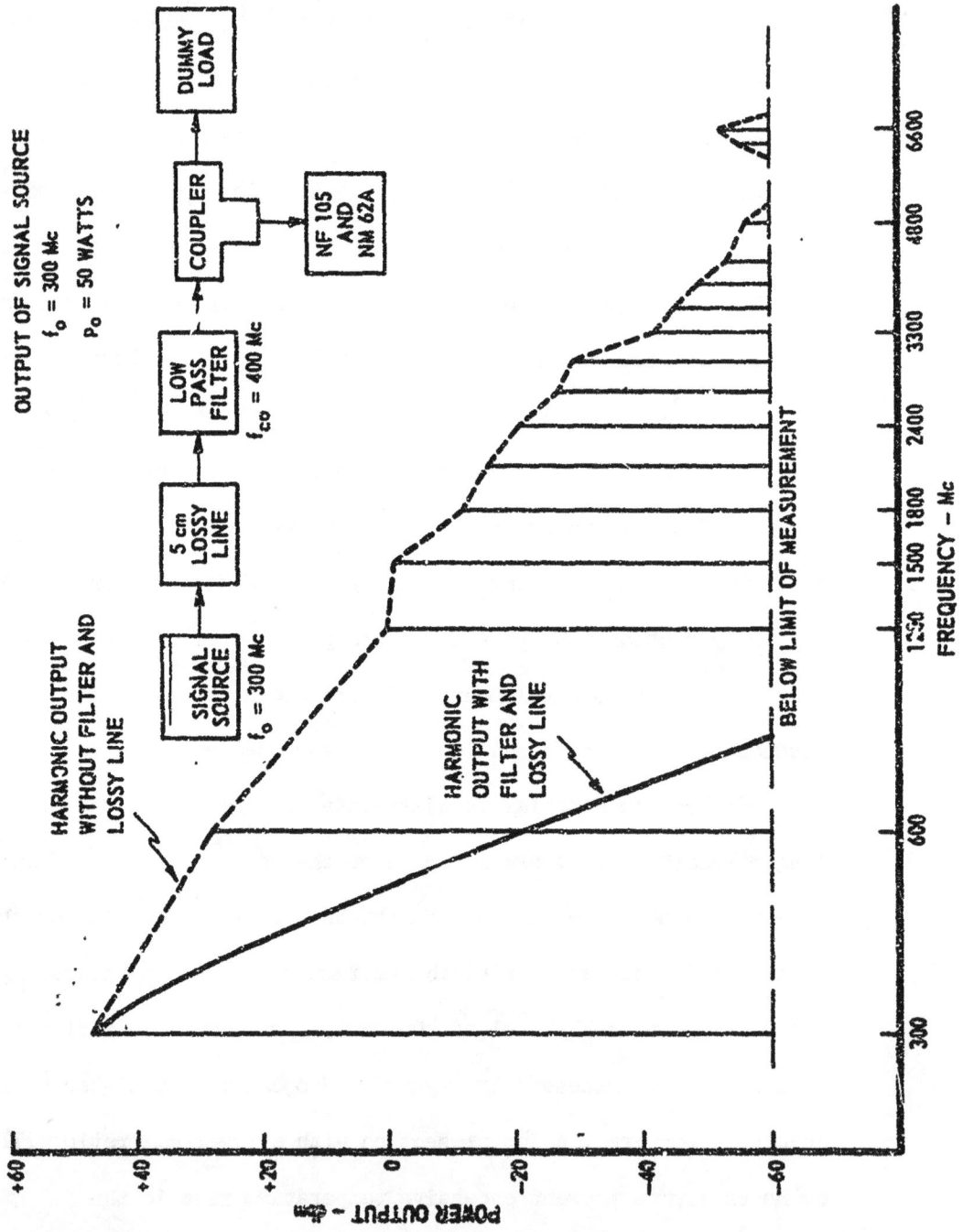

Figure 39. Harmonic Output of Signal Source.

much below 2:1 are not desirable, the limitation on power handling capability would seem to be the temperature rise in the dielectric rather than the breakdown potential.

2.1.10 Lossy Filter Construction

Once a particular filter design has been chosen, the actual construction of the filter shell and the preparation of the absorbing material are straightforward procedures. The filters built for laboratory testing were designed to be compatible with 50 ohm impedance measuring equipment.

The filter body shell was constructed from standard 5/8 inch O. D. brass tubing with 1/32 inch walls. The 5/8 inch tubing mates conveniently with a standard type N cable connector which is easily attached if the nut of a standard type N connector is butt soldered to each end of the tubing. If high powered operation of the filter is contemplated, cooling fins should be soldered to the shell before the ends are attached. Care should be taken when calculating the total length of the center conductor to be used to insure the proper mating of the center pins of the connectors when the filter is connected into a system.

Molds are necessary to hold the absorbing mixture in place until it hardens and to properly position the center conductor. The diagram of Figure 40 shows the construction of the Teflon molds which were used in pouring the filters used for testing. When pouring the lossy mixture, the mold was overfilled to allow for any shrinkage and the excess machined off after hardening.

88

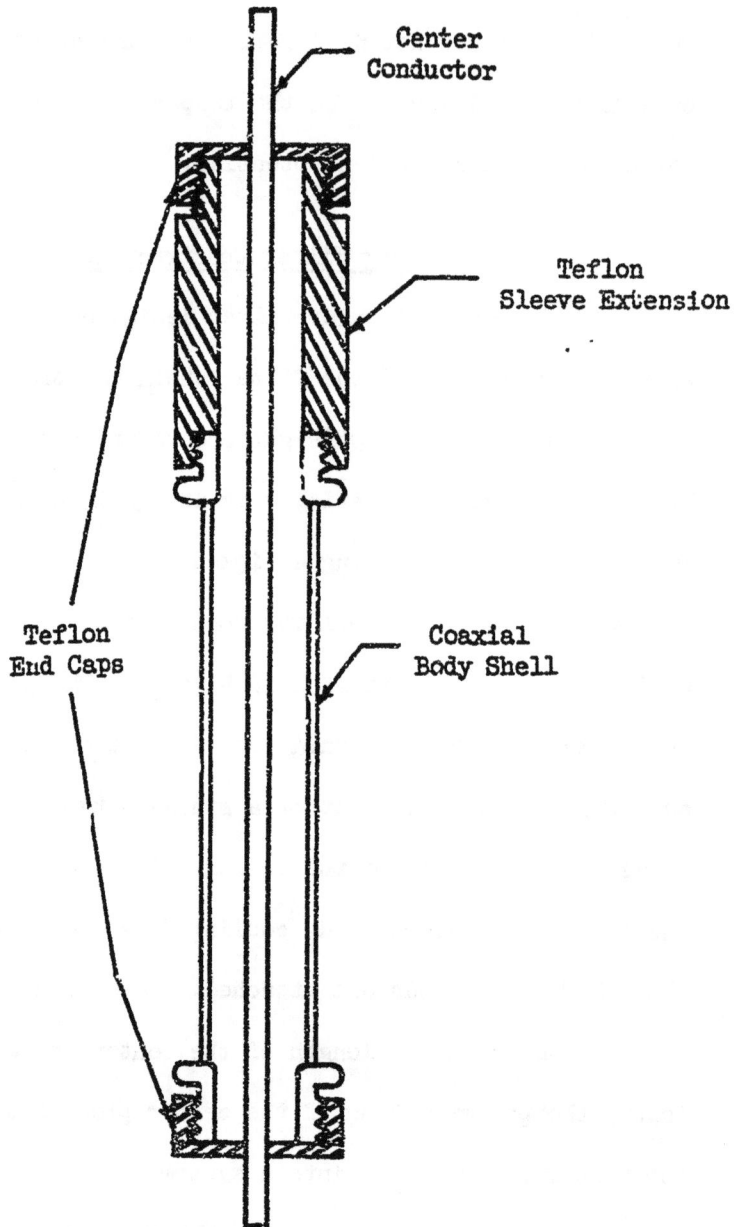

Center
Conductor

Teflon
Sleeve Extension

Teflon
End Caps

Coaxial
Body Shell

Figure 40. Mold Construction.

The center conductor, outer shell, and molds should be prepared
before beginning the preparation of the absorbing material, since
the epoxy begins to set up in twenty to thirty minutes. The amount
of absorbing powder to be used will depend upon the mix ratio desired
and the required volume to be filled. The total amount that should
be prepared can best be estimated after a few trial runs; the weight
ratio of absorbing powder with the particular powder being used having
been chosen from design considerations. The epoxy binder is a mixture of epoxy, hardener, and a debubbling agent. Epon Resin 828
manufactured by Shell Chemical Company was used as the epoxy and the
hardener was diethylene triamine marketed by Magnolia Plastics, Inc.
The debubbling agent was Plastolein 9051 DNHZ, which was obtained
from Emery Industries, Inc. Other epoxies, with appropriate
hardening agents, can be used if the Shell Epon 828 is not available.

To prepare the binder, first estimate the weight of absorbing
powder required and weigh out an approximate amount of epoxy for the
mix ratio desired. Call this amount of epoxy X grams. Add 0.1X
grams of hardener to the epoxy and add 0.11X grams of plastolein to
the epoxy and hardener. The total binder mixture is now

$$X + 0.1X + 0.11X = 1.21X \text{ grams} \tag{29}$$

The proper weight of absorbing powder to be used is equal to the mix
ratio, M, times the weight of binder.

The epoxy, hardener, and plastolein should be mixed thoroughly

90

before the powder is added. A simple mixing aid was constructed by making a brass paddle on a 1/4" brass rod and rotating the paddle by the use of a small hand drill. The speed of the drill motor was adjusted by varying the supply voltage with an autotransformer.

Once the binder is thoroughly mixed, the powder should be slowly added to the mixture while continuously blending the two. A uniform mixture is necessary to obtain reproducible attenuation characteristics for the filter.

Once thoroughly blended, the epoxy-powder mixture is ready to be added to the filter section. The ease with which the mixture pours into the cavity depends upon the viscosity of the mixture. The viscosity is established by the mix ratio and the density of the powder. For instance, a 6:1 mix of carbonyl iron powder pours readily but a 4:1 mix of ferrite powder or mill scale is as high a mix ratio as can be easily poured. A flexible plastic bottle with a nozzle was found to be useful in getting the iron-epoxy mixture into hard to reach places.

Due to the viscous nature of the epoxy-powder mixture, there is a tendency for air bubbles, which are produced in mixing, to be trapped inside the hardened material. These bubbles can upset the impedance characteristics of the dielectric material and alter the attenuation characteristics. The removal of these bubbles is facilitated by vibrating the filled line section during the curing stage. Slight heating of the mixture aids in driving off the bubbles but this technique should be used with care because heat also speeds the reaction time of the epoxy and the bubbles can be trapped before they have a chance to rise to the top of the dielectric mixture. If a vacuum system is available, the filter can be placed under a slight vacuum to speed the removal of bubbles but the vacuum does not significantly speed the reaction time of the epoxy.

2.2 HF Filters with Dielectric Loading

Coaxial resonators are commonly used as filter elements in the microwave region where a typical filter might consist of several quarter wave coaxial cavities with suitable coupling means between the resonators. As the design frequency of such filters is decreased, the physical size of the resonators increases until, at frequencies in the VHF and HF region the necessary resonators become excessively large. The helical resonator is one technique commonly employed to reduce the size of resonators used in the VHF region. The helical resonator uses a center conductor in the form of a helix to increase the electrical length of the resonator.

The low losses and low propagation velocity at HF of TEM waves in the dielectric materials used in the construction of lossy filters suggests the possibility of further reduction in the size of such resonators to the extent that HF operation becomes feasible.

2.2.1 Construction of HF Filters

Several helical resonators were constructed to evalute the amount of reduction in physical dimensions that could be obtained by the use of dielectric loading. All of the resonators were designed to resonate with an air dielectric in the VHF range and to resonate in the HF range when filled with a dielectric material.

A half wave resonator was constructed in the form shown in Figure 41. The resonant frequency with an air dielectric was 107 Mc and the insertion loss was 0.8 db. When the dielectric space was filled with SF grade carbonyl iron powder, the resonant frequency decreased to 23 Mc and the insertion loss increased to 8.5 db. Part of this insertion loss was caused by mismatch loss at the input and output terminals of the resonator, and the remaining loss was due to dissipation in the dielectric material.

Another HF filter was constructed in the form indicated by the sketch of Figure 42. The resonators are of the top loaded quarter-wave type described in a recent Stanford Research Laboratory report.[4] The filter was tested both with an air dielectric and with the dielectric space-filled with a carbonyl iron material. The epoxy binder used with the SF grade carbonyl iron was Emerson and Cuming's Sytcast No TPM-2, which has a smaller dielectric loss tangent than that of the epoxy commonly used.

The response obtained for the filter filled with the slow velocity dielectric material is compared in Figure 43 with that obtained for the same resonator using an air dielectric. The

TOP VIEW

CUTAWAY SIDE VIEW

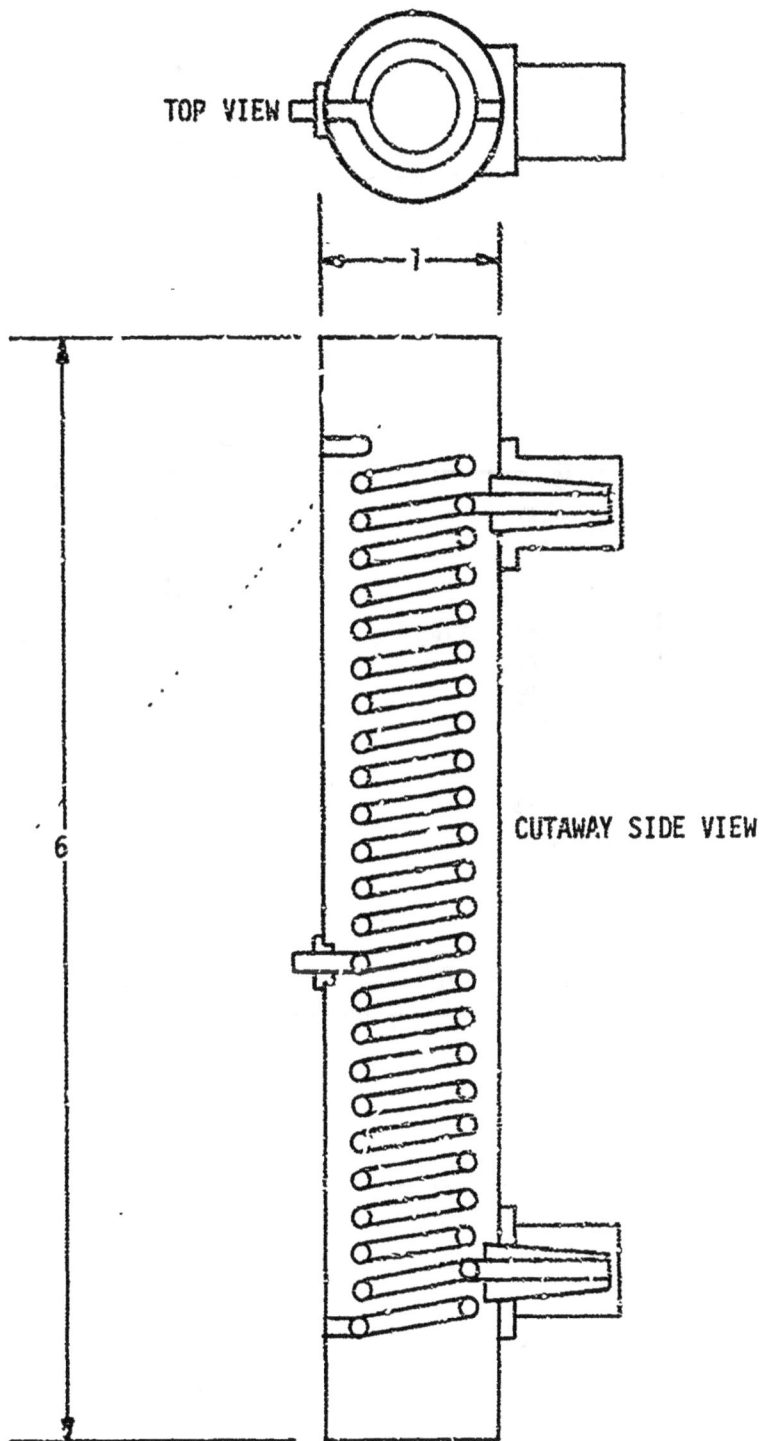

Figure 41. Half-Wave Resonator.

TOP VIEW WITH COVER REMOVED

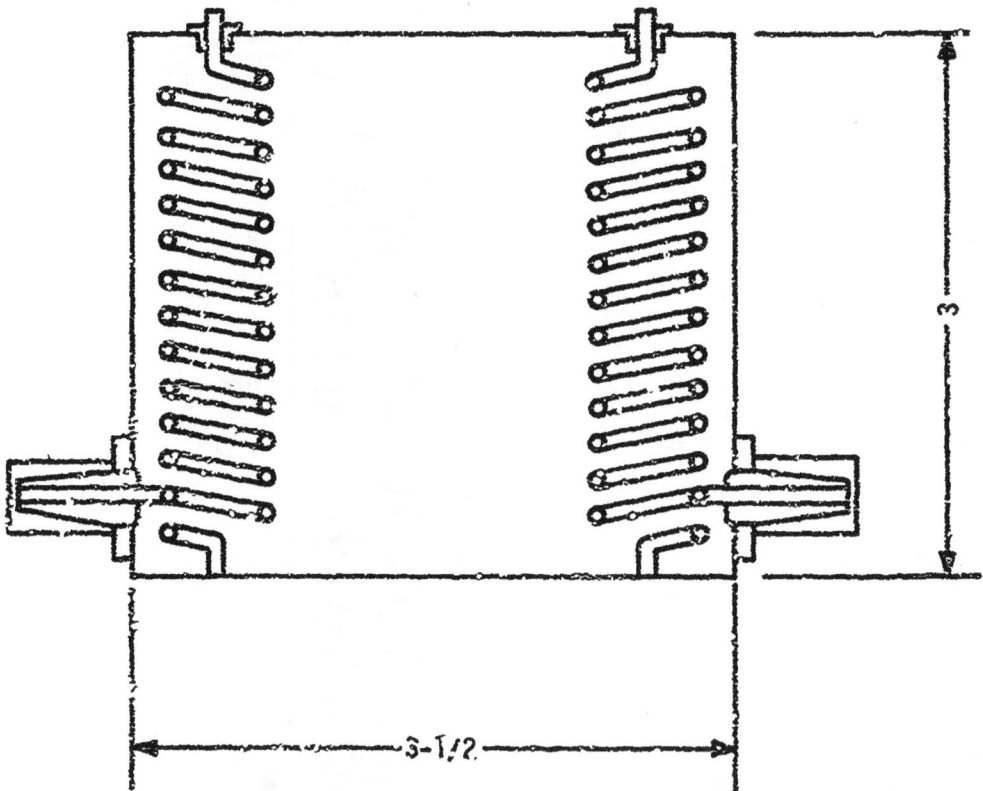

CUTAWAY SIDE VIEW

Figure 42. RF Filter.

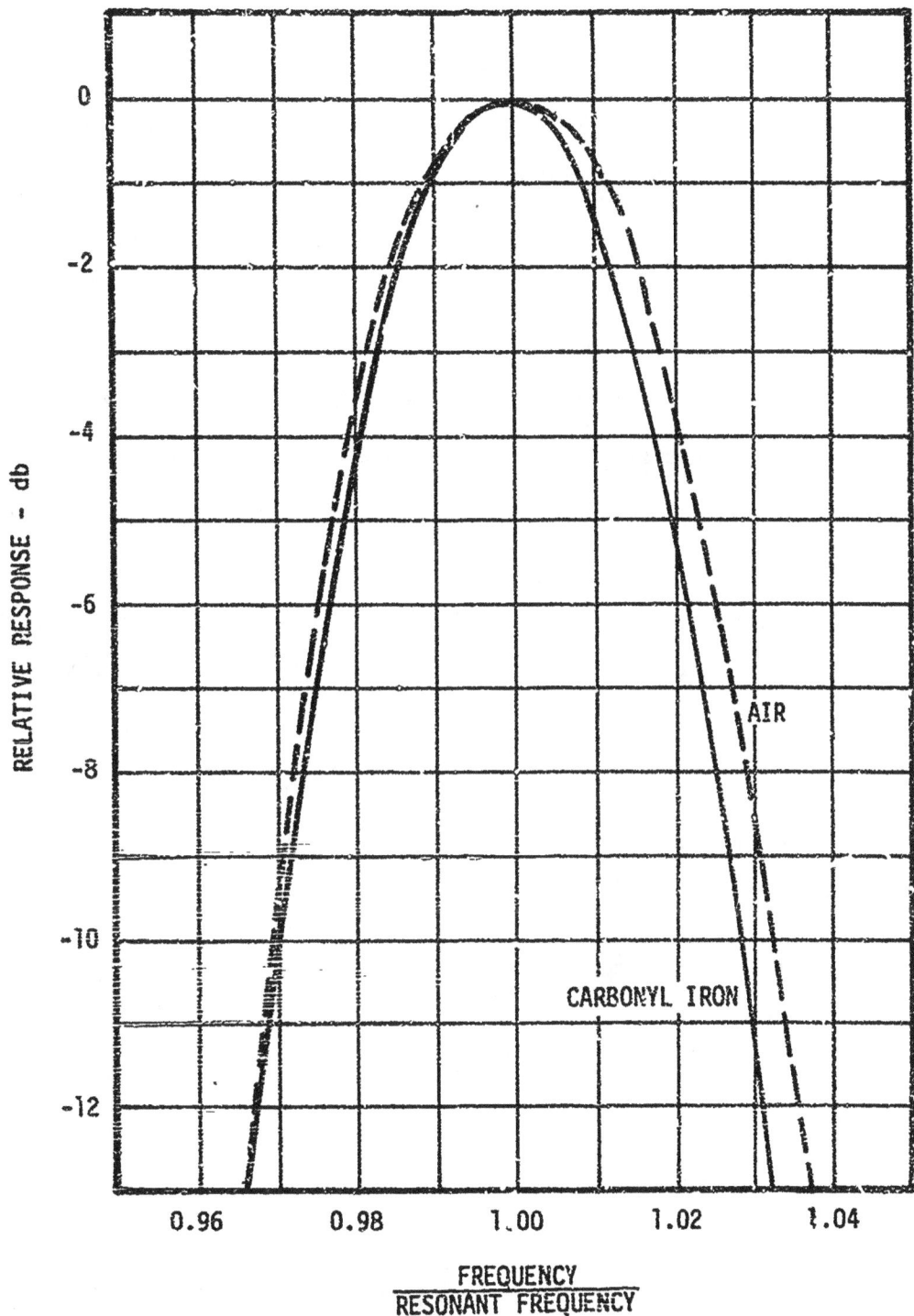

Figure 43.   Normalized Response of HF Filter.

actual resonant frequency of the air dielectric resonator is
approximately six times that of the iron-filled resonator and the
normalized selectivity curves are used to illustrate the change in
shape factor which occurs when the iron dielectric material is
added. The passband insertion loss increased to 5.8 db when the
resonator was filled with lossy dielectric. Again, a portion of
the insertion loss may be attributed to the lack of proper impedance matching when the resonator was filled with the iron dielectric,
and the remaining loss, to dissipation losses in the iron dielectric.
In an effort to reduce the passband insertion loss, another filter
was constructed using a closer spacing between the two helices.
With an air dielectric, the center frequency of the filter was 52
Mc. To avoid an over-coupled response for the filter, a vertical
copper shield was placed between the helices. The coefficient of
coupling was then adjusted by increasing the dimensions of a slot
in the copper shield until the desired flat topped response was
obtained. When powdered iron was added to the dielectric space,
the tuned frequency dropped to 8.5 Mc and the shape of the response
indicated that the coupling coefficient was only slightly greater
than critical even with the copper shield completely removed. The
frequency response of the iron filled filter is shown in Figure 44
which shows the desired narrow response at 8.5 Mc and a broad
spurious passband in the region of 42 Mc. The response in 42 Mc
region is not suppressed by losses in the carbonyl iron since the
dissipation loss at 42 Mc for carbonyl iron is small. The addition

97

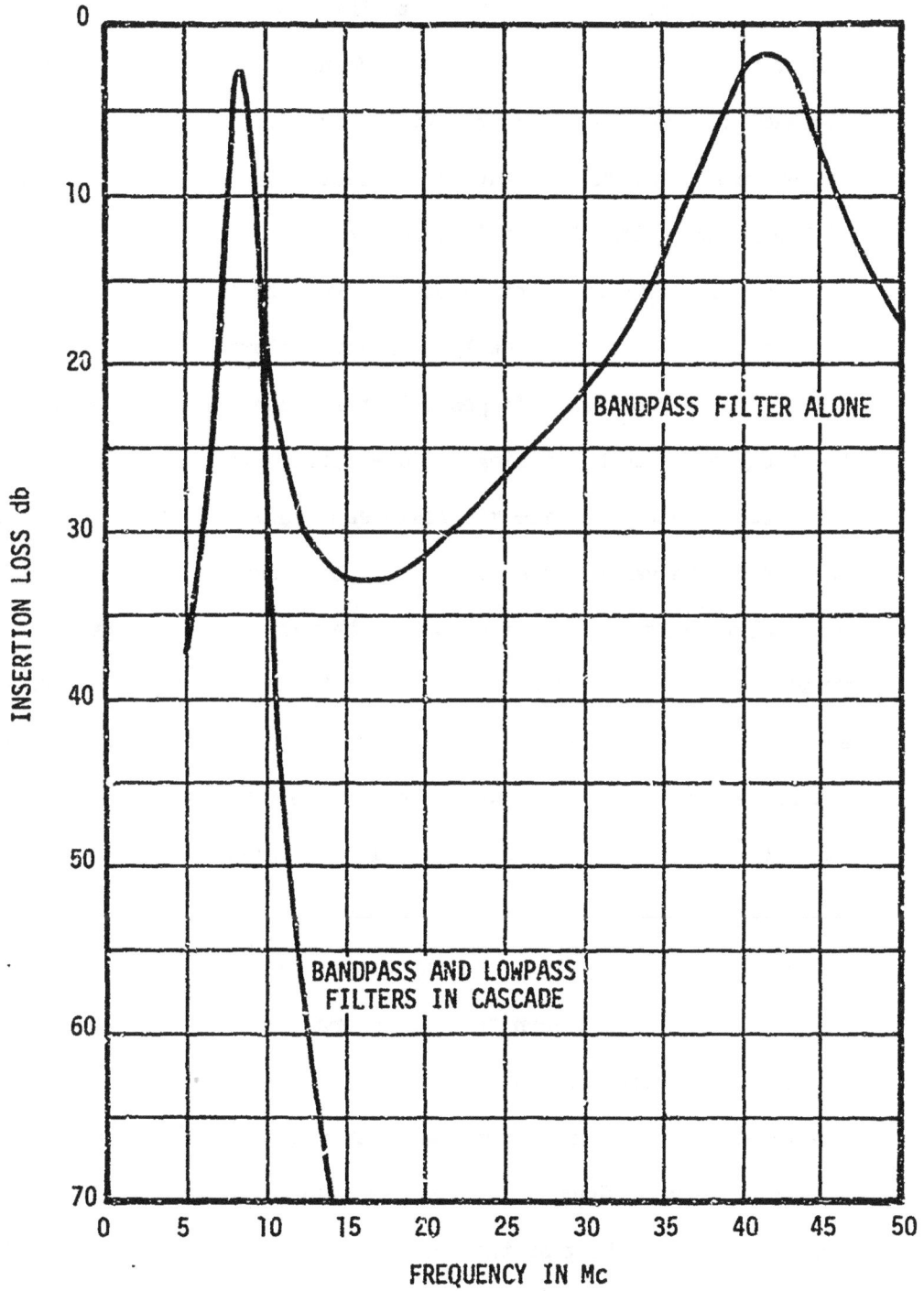

Figure 44. Response of Iron Filled Bandpass Filter.

of a low-pass filter in cascade with the iron filled resonator can effectively suppress the spurious passband. The response of the cascaded combination of an iron filled resonator and a low-pass filter is also shown in Figure 44, illustrating a band-pass response with uniformly high stop-band rejection.

Several different ferrite powders were tested to determine their suitability for use as low velocity materials for reducing the size of resonators. In general, the reduction in resonant frequency obtained by filling the dielectric space of a helical resonator with the ferrite materials was consistent with that estimated from the values of μ_r and ϵ_r of these materials. The ferrite material with lowest losses at the 10 Mc test frequency was Q-3 ferrite. One difficulty encountered with the ferrite filled filter employing side by side helices was the inability to obtain adequate mutual inductance coupling when the ferrite powder was added to the resonator. One solution to the coupling problem is to add an external coupling capacitor between the high impedance ends of the resonators. Adjustment of the capacitor then permits any desired degree of coupling between the resonators to be obtained. Using a two section ferrite filled filter similar to that described in Figure 42, passband insertion losses of approximately 2 db were measured. The passband insertion was reduced to this value by increasing the loading of the resonators with an adjustment of the input and output tap points on the helices. In general, to minimize passband insertion losses, the impedance transformation should be set so that the relative amount of loading of the resonator due to the reflected load impedance is large with respect to the loading due to dielectric losses in the resonator itself. As long as this loading condition is met, the tap point required for a given percent power transmitted is given by

$$N = \left[\left(\frac{Z_R}{Z_L} \right) \left(\frac{100 - \% \text{ Power}}{\% \text{ Power}} \right) \right]^{1/2} \qquad , \qquad (30)$$

where N is the ratio of the helix turns at the tap point to the total number of turns in the helix.

2.2.2 Effects of Static Magnetic Fields

Some tests of the effect of a static magnetic field on the tuned frequency of a ferrite filled filter were made. With zero field applied, the tuned frequency was set to 10 Mc. A static magnetic field was applied to the filter and the frequency increased to 15 Mc. Although some small changes in the shape of the selectivity curve were noted, these do not appear to pose a serious drawback to the use of such a device as an electrically tunable bandpass filter.

A more serious problem is the large magnetic field required to produce significant changes in the tuned frequency of the resonator. In the tests conducted, the external magnetic field was supplied by a large permanent magnet and the applied field was varied by moving the magnet. In order to provide rapid tuning of the resonator, an electromagnet would be necessary and the power dissipated in this magnet might become excessive. It is possible that by proper mechanical arrangement of the tuning magnet with respect to the resonator, the air gap between the magnet and the dielectric material could be reduced. If such an arrangement could be realized, a substantial reduction in the number of ampere-turns required to saturate the ferrite dielectric material might be obtained.

2.2.3 High Power Operation of Iron and Ferrite Loaded Filters

The ability to tune a ferrite filled resonator magnetically is an attractive property for some applications, but it also suggests that changes in the resonator may take place when a large RF magnetic field is present. A ferrite filled filter was terminated with a 50 ohm dummy load and driven at several different RF power levels to examine the effect of power level on the characteristics of the filter. The filter was constructed of two concentrically mounted helices with the input helix being of slightly smaller diameter than the output helix. The concentric construction was used to increase the mutual coupling between the helices to the point where no external capacitive coupling between the input and output resonators was needed. The wide range of power levels needed to carry out the tests were supplied by an HF power amplifier. The amplifier was driven by a low power signal generator and was capable of delivering an output of 100 watts over the HF range. Appendix D gives a discussion of the design and construction of the power amplifier. The curves of Figure 45 show the response of the cavity response at various RF power levels. As the CW power level is increased from 10^{-3} watts to 2 watts and then to 20 watts, there is a definite variation in the insertion loss, the tuned frequency, and the skirt selectivity of the ferrite filled resonator. The time scale on which the observed effects take place indicate that the change in filter characteristics is due, at least in part, to a shift in the temperature of the ferrite. For example, when the resonator was driven from a 20 watt level, the initial insertion loss was typically 1 db. However, the insertion loss was observed to increase slowly with time as the temperature of the ferrite in the filter increased.

The data shown in Figure 45 was obtained with power being applied to the cavity only a short time at each point on the curve in order to avoid the shift in resonant frequency as a function of temperature. Since the measurement at each frequency was taken quickly, the change in the response of the cavity with increasing power level was probably due to the RF magnetic field as opposed to variations resulting from increasing temperature. To definitely isolate temperature effects, the resonator was tested in a pulsed power mode. The test pulse was obtained by setting the level of the low power signal to drive the power to the desired peak RF output and then pulse modulating the low power signal generator at a duty cycle sufficiently low to limit the average power delivered to the cavity to 10^{-3} watts. The results of these tests, shown in

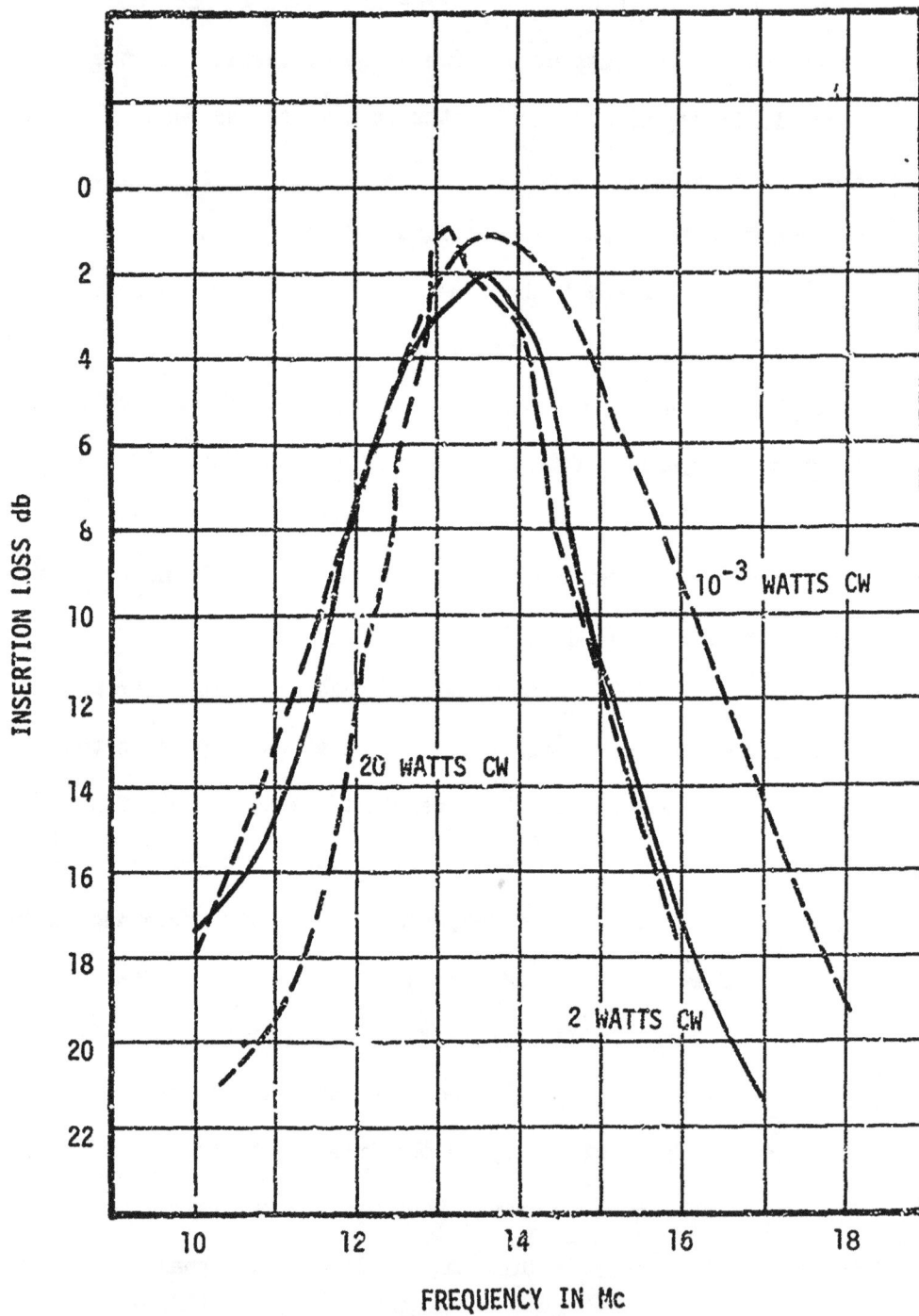

Figure 45. Effect of CW Power Level on Ferrite Filter.

Figure 46, indicate that the response characteristics of the resonator are sensitive to the level of the RF magnetic field.

Additional tests were made on a resonator filled with carbonyl iron. The response characteristics of the resonator as a function of power level are shown in Figure 47. As indicated by the data, there are fewer changes in the insertion loss, skirt selectivity and tuned frequency with the carbonyl iron filled resonator compared with that of the ferrite filled resonator.

To insure an accurate measurement of the insertion loss of the carbonyl iron filled cavity, it was necessary to use a fixed pad at the input of the cavity to establish a 50 ohm impedance level. The 10^{-3} watt and 2 watt power measurements were CW operation whereas the 20 watt data was obtained using a pulse mode. It was necessary to operate in a pulse mode at the 20 watt level to maintain a low average power so that the 50 ohm fixed pad would not be damaged. To test for temperature effects, the resonator was operated at a 100 watt CW power level, without a pad, for an extended time and no significant shift in the resonant frequency as a function of temperature was observed. Since the resonant frequency of the cavity is stable at high CW power levels, which was not the case with the ferrite filled cavity, the 20 watt pulse power test is equivalent to a 20 watt CW test as far as insertion loss measurement is concerned.

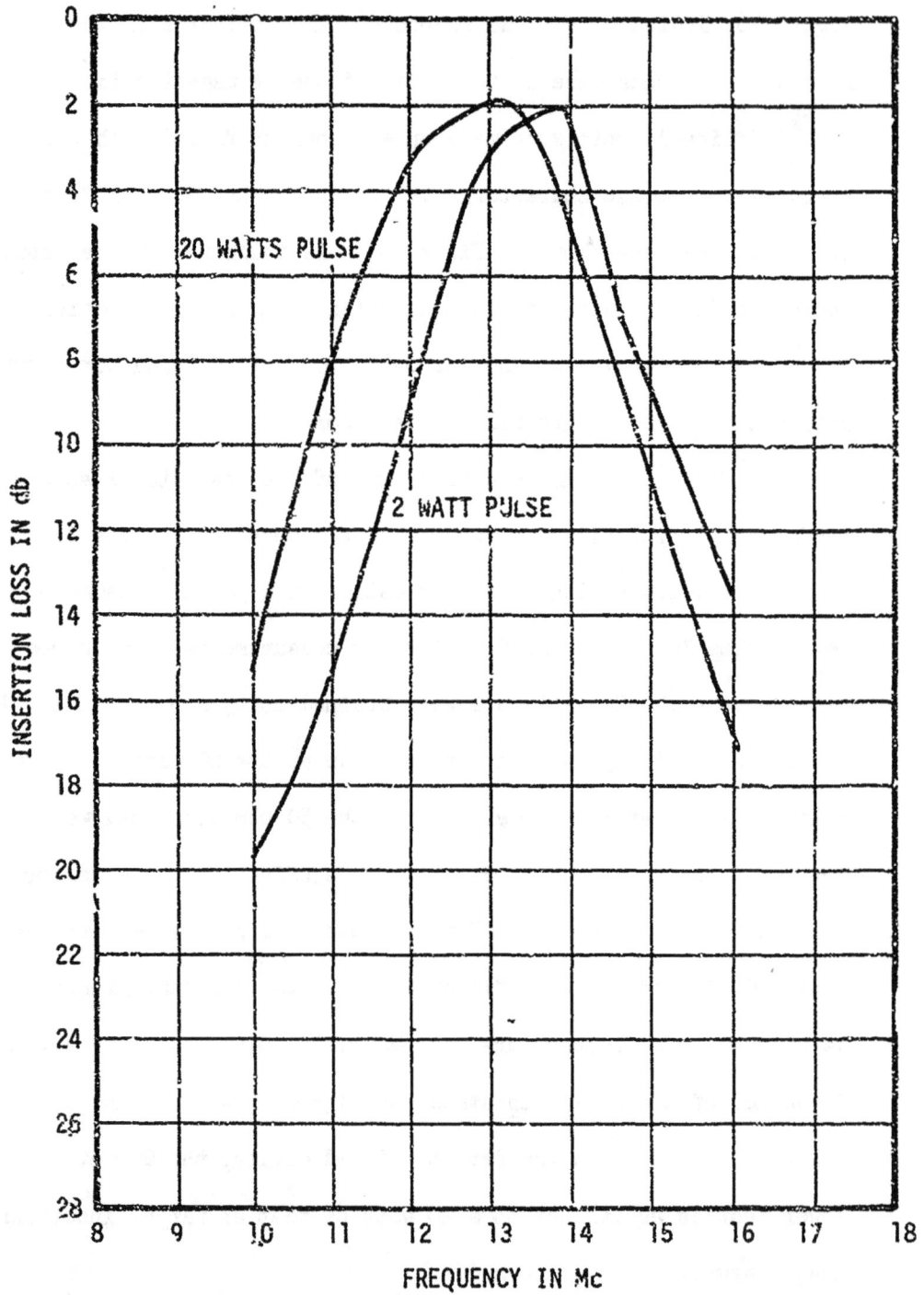

Figure 46. Effect of Pulse Power Level on Ferrite Filter.

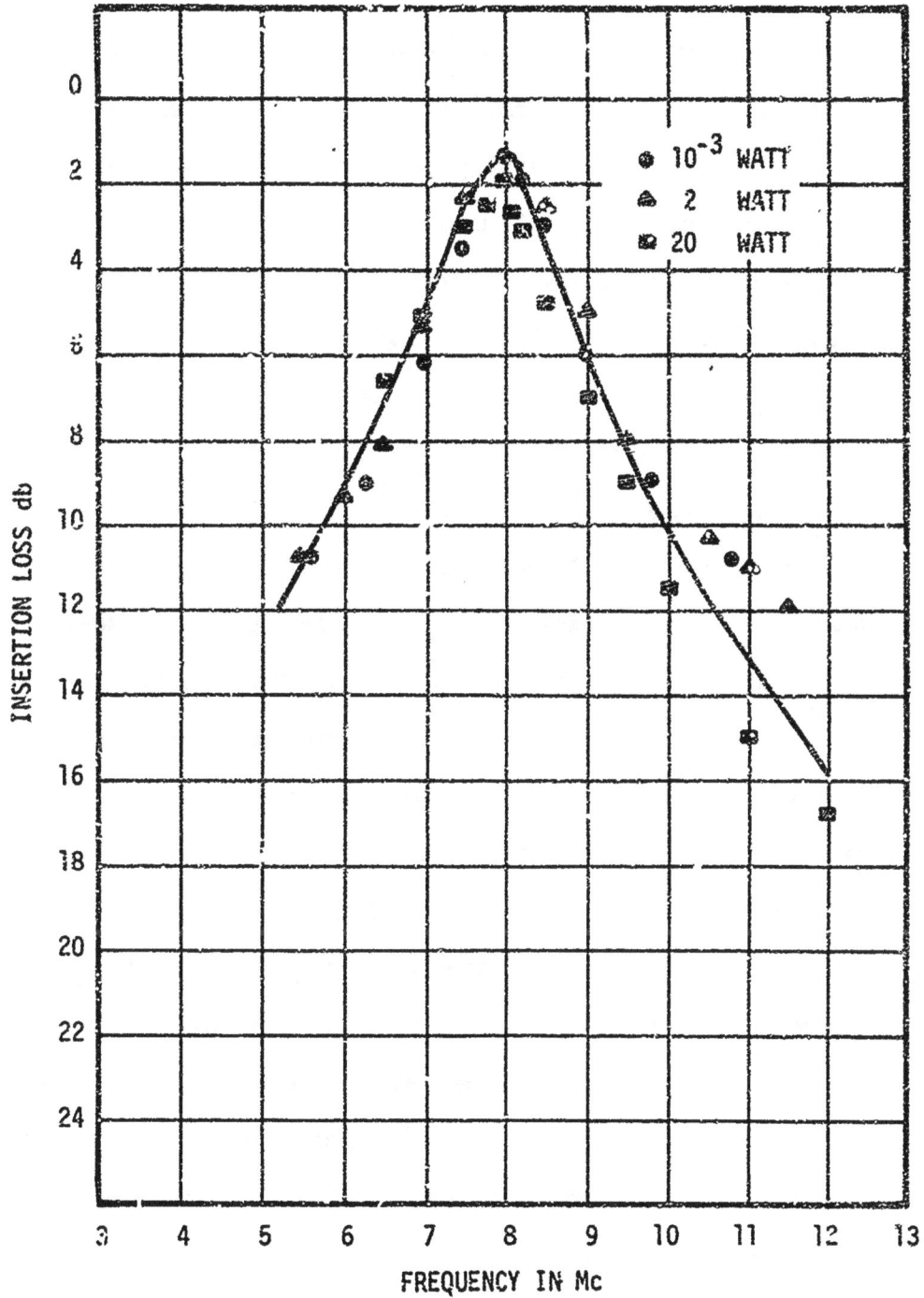

Figure 47.   Effect of CW Power Level on Carbonyl Iron Filter.

106

2.3 Shielding Effectiveness Measurements

Relative shielding effectiveness measurements of several dielectric materials were made by measuring the difference in coupling between a transmitting and receiving loop with and without the dielectric material in place between the two loops. Since the materials were in powdered form, a special container was constructed to fit over the receiving loop to allow the dielectric material to be easily changed. The construction of the container is shown in the sketch of Figure 48. The container consists of two polystyrene boxes of slightly different dimensions with the smaller box nested inside the larger. The transmission between the loops was measured with no dielectric material in the space between the boxes. The air space between the boxes was then filled with the material being tested. The box was inverted and bolted in place over one of the loops and the transmission between the two loops was measured. The difference in transmission between the two sets of measurements was taken to be the shielding effectiveness of the material in the air space between the two boxes. In order to remove the effects of reflections from the surroundings, the two loops were placed inside a large metal box which had been lined with absorbing material. The absorbing material reduced the effects of wall reflections above approximately 700 Mc.

Some typical transmission data obtained for a carbonyl iron material is shown in the curve of Figure 49. The results obtained at frequencies below 1 Gc indicate a lower attenuation with the

Double Walled
Plexiglas Box

Loop
Antennas

Ground
Plane

Wooden
Support

Type N
Connectors

Lossy
Dielectric
Powder

Ground Plane
Extension

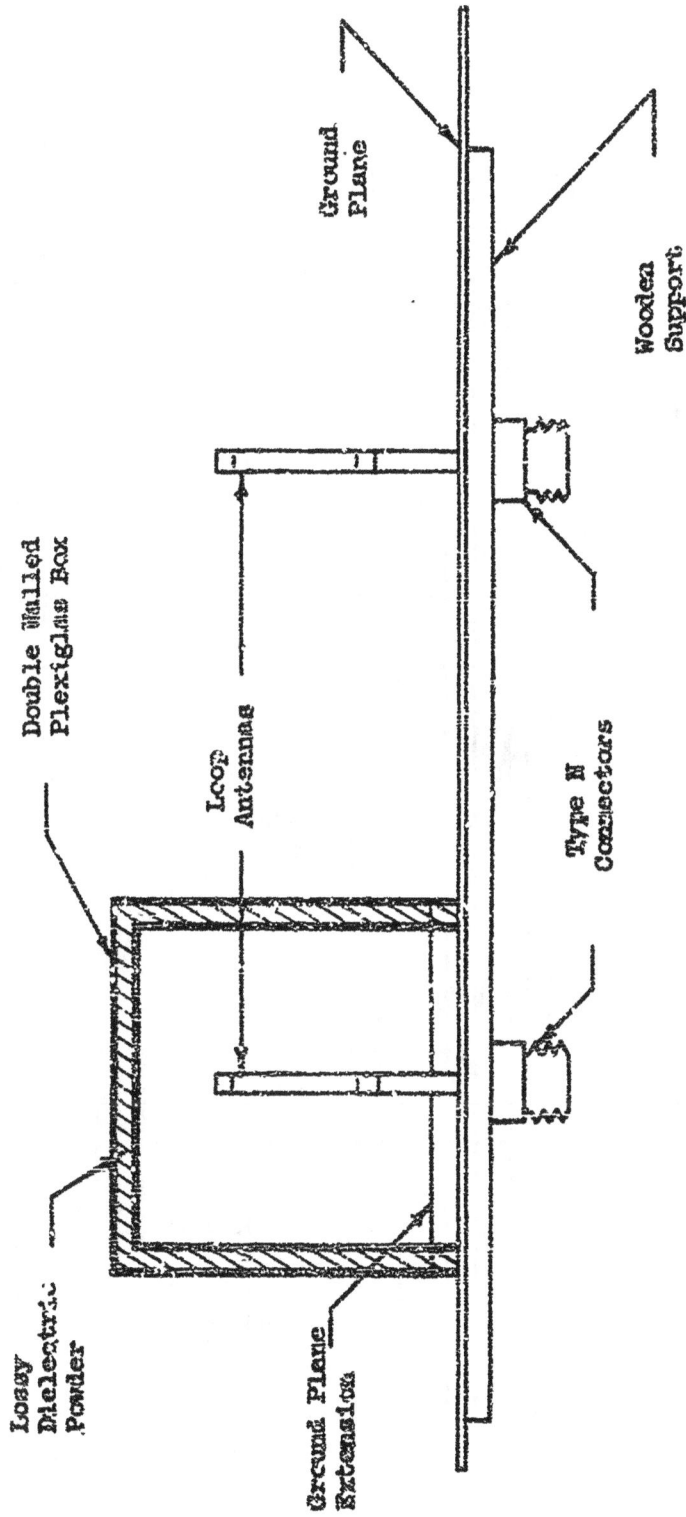

Figure 48. Shielding Test Assembly

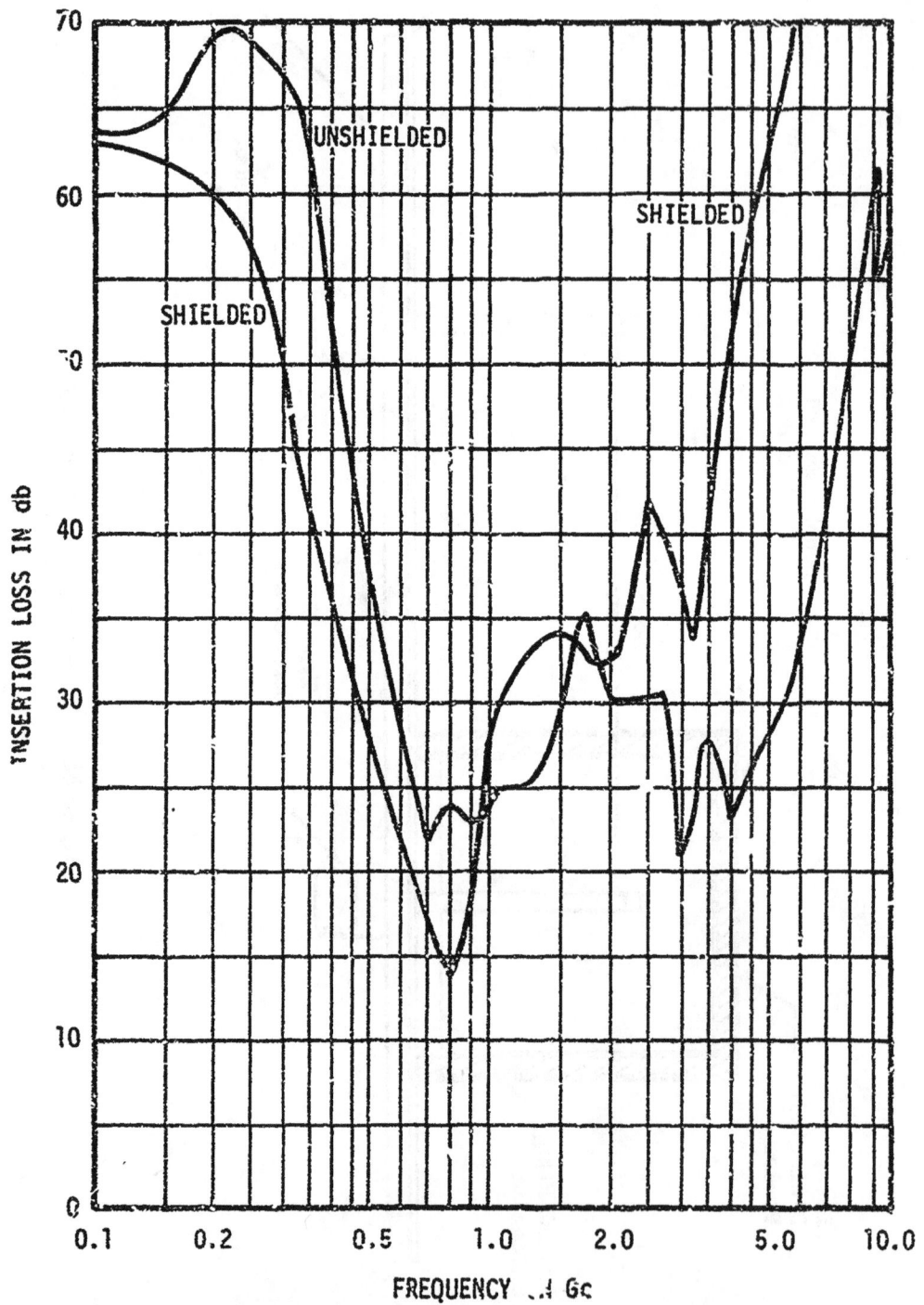

Figure 49. Shielding Effectiveness Test Results.

shielding material in place than with it removed. A possible explanation for such a result is the change in loop current distribution that occurs when the test material is in place. The change in the current distribution alters the input impedance of the loop in such a manner as to improve the impedance match between the signal source and the loop. The increased power delivered to the loop could then account for the apparent negative shielding effectiveness. Another contributing factor is the lack of effectiveness of the absorbing material over much of the frequency range below 1 Gc which causes the pattern of reflections inside the box to be modified by the placement of the shielding material under test. As a result of these factors, the data obtained on shielding effectiveness below 1 Gc is unreliable. At frequencies above 1 Gc, the absorbing material on the walls becomes increasingly effective and the effect of the shielding material on the current distribution on the loop diminishes rapidly. As a result, the data in the region above 1 Gc is more representative of the true shielding effectiveness of the dielectric material.

The data obtained for both mill scale and ferrite powders did not indicate appreciable shielding action over the range of measurement, since the loss curves for these two materials followed a similar pattern to the calibration curve which was taken with no material between the loops.

Additional measurements were made to examine the shielding effectiveness of a carbonyl iron lossy material at X-band. The test specimen used was a 1/4 inch thick slab of 6:1 iron to epoxy material. The slab was placed across the mouth of an X-band horn and the change in output of the detector connected to another horn was recorded. The change in the VSWR at the input to the transmitting horn when the slab was placed in position was also noted. The general measurement setup is shown in Figure 50.

The results of these tests are summarized in Figure 51 which shows the insertion loss and VSWR variation in the X-band range. There is a rising attenuation characteristic with increasing frequency similar to that observed for coaxial systems. The VSWR measurements show an average VSWR of approximately 2:5:1, indicating that most of the insertion loss is produced by dielectric losses in the slab rather than by reflection from the impedance discontinuity presented by the slab at the mouth of the horn.

2.4 Rapid Insertion Loss Measurements

A large number of insertion loss measurements were necessary to determine the loss characteristics of the several dielectric materials investigated. In order to reduce the time required to obtain the necessary attenuation data, some effort was devoted to the development of an automatic loss measuring system which would plot directly the loss versus frequency characteristic of the device under test. One technique used in making these automatic measurements is shown in Figure 52. Such an arrangement can

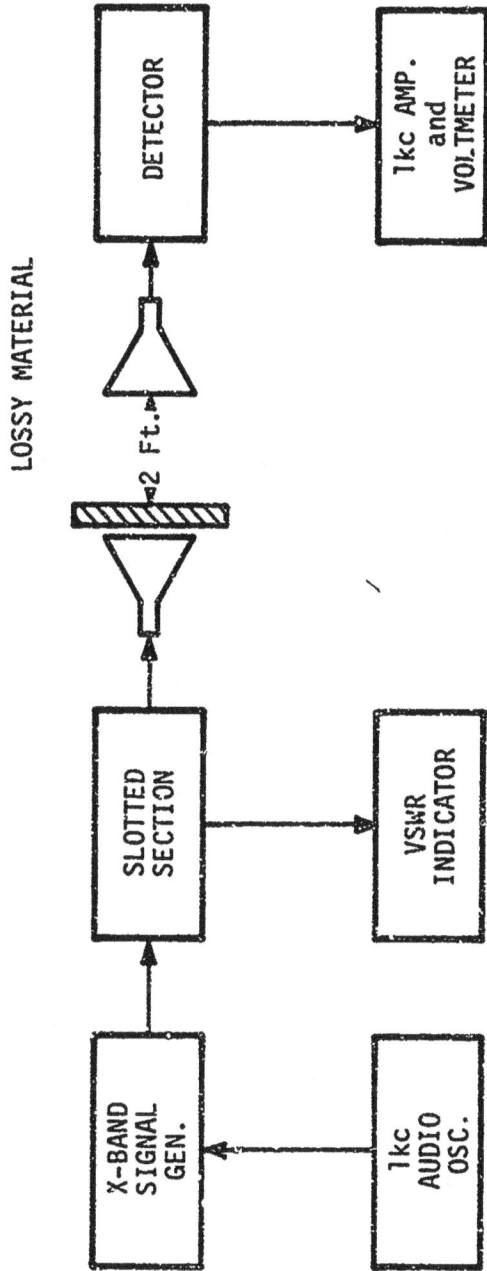

Figure 50. X Band Shielding Test Assembly.

112

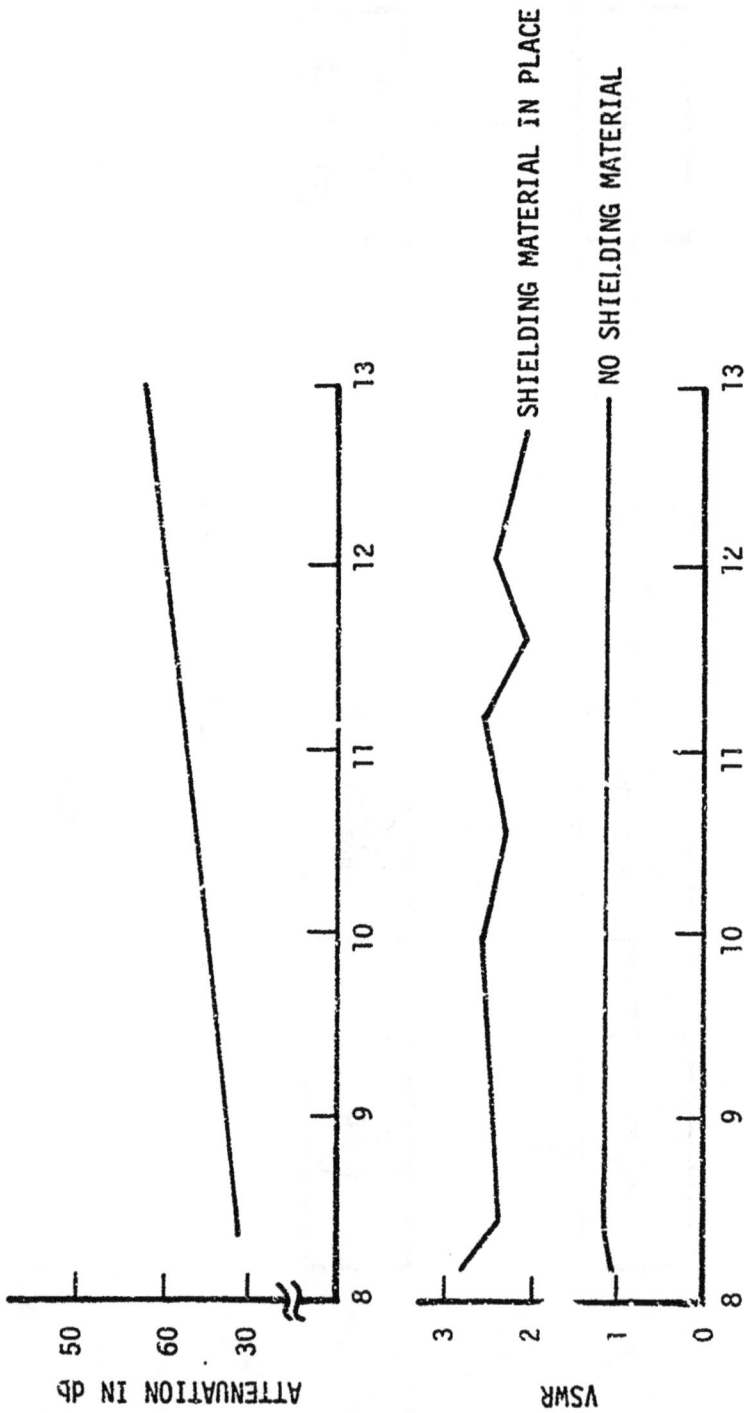

Figure 51. X Band Shielding Loss.

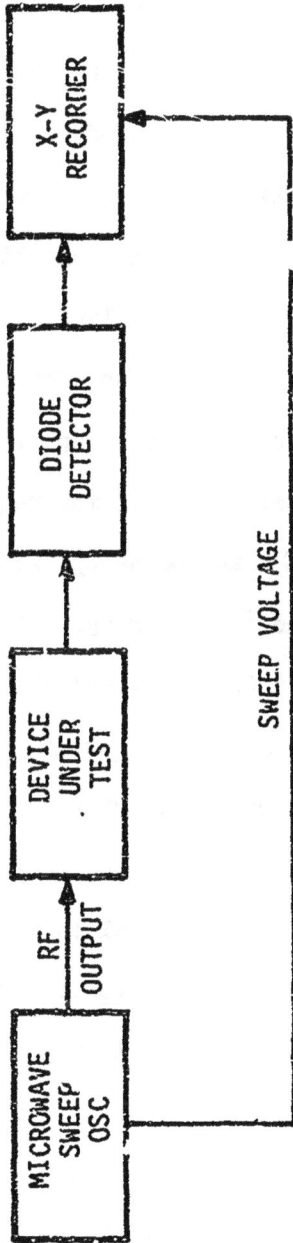

Figure 52. Rapid Insertion Loss Measurement System.

114

rapidly produce attenuation data when the loss of the device being tested does not exceed 30 to 40 db. For attenuations higher than this value, the loss in sensitivity of the diode detector at low signal levels seriously limits the sensitivity of the overall measurement system. To overcome the sensitivity limitation, the superheterodyne detection system outlined in Figure 53 was constructed. For such a system to be effective, the frequency of the local oscillator signal must be swept in synchronism with the frequency of the swept test signal. The required local oscillator signal was obtained from a 2-4 Gc sweep oscillator. For test signals in the 4-8 Gc and 8-12 Gc ranges, the local oscillator sweeps over its basic 2-4 Gc range, and the 30 Mc IF signal is generated by harmonic conversion in the mixer. The linearity of the control voltage-frequency characteristic of the sweep oscillators used was not sufficient to provide the necessary tracking to maintain a 30 Mc offset between the local oscillator signal and the frequency of the test signal. An AFC loop was installed on the local oscillator to provide the necessary tracking accuracy. To obtain the AFC action, the output of the signal IF amplifier is fed to a limiter and a discriminator tuned to 30 Mc. When the local oscillator frequency is properly set, no error is developed at the discriminator output. However, any lack of tracking of the local oscillator signal produces a DC output voltage from the discriminator. This DC voltage is superimposed on the local oscillator sweep voltage in the proper polarity to return the IF

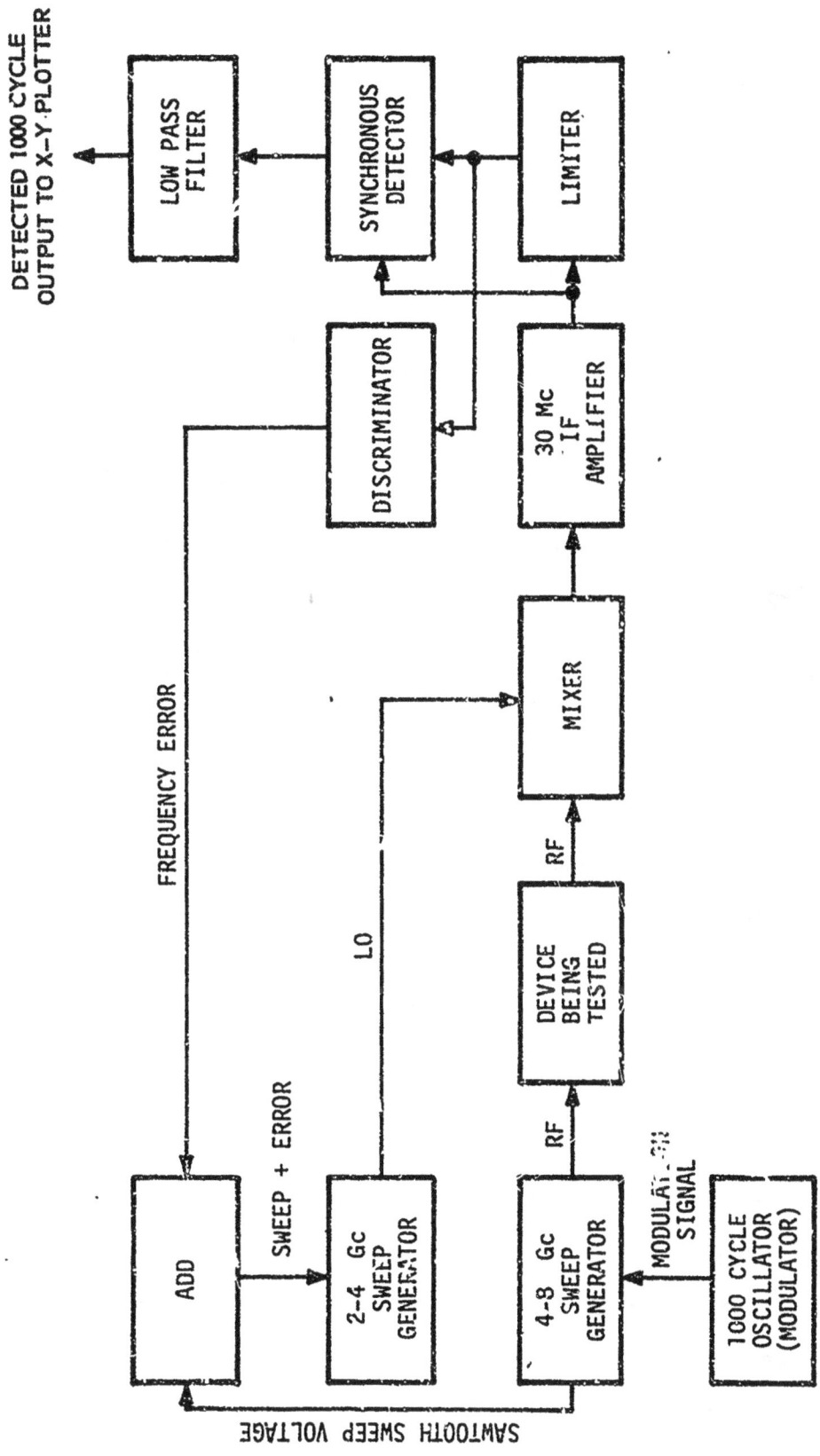

Figure 53. Superheterodyne Measurement System.

signal to 30 Mc. Sufficient loop gain is present to maintain the
IF frequency under normal sweep conditions in the range 30 Mc ±
0.25 Mc.

A certain minimum signal level is necessary at the output of
the IF amplifier to maintain linear operation of the 30 Mc diode
detector. If the linear range of the measuring system must be 60
db, then the IF amplifier must be capable of providing linear
amplification at a level 60 db above the minimum level required
for the diode detector. Conventional IF amplifiers are not
capable of supplying the high level signals necessary to meet the
60 db dynamic range requirement. A straightforward solution to
the dynamic range problem is the use of a logarithmic IF amplifier
to compress the range of signal levels that must be supplied at
the output of the IF amplifier. However, at the time of construction of the measuring system, no logarithmic amplifier was available.
Instead, the 60 db dynamic range was obtained by replacing the 30
Mc diode detector with a synchronous detector. Since the synchronous
detector requires no minimum signal to maintain linear operation,
the necessary dynamic range can be obtained by operating the
detector at signal levels in the range between 1 millivolt and 1
volt. Such levels are well within the linear operating range of the
IF amplifier. The reference signal required to switch the
synchronous detector is supplied from the constant level output of
the limiter. The phase shift versus frequency curve of the limiter
is sufficiently flat over the ±250 kc range through which the IF

signal varies that no significant charge in the phase occurs in
the limiter output signal. Consequently, a fixed adjustment of
the phase of the synchronous detector reference signal is
sufficient to insure the proper phase of the reference signal over
the ± 250 kc swing of the IF signal. The 1 kc detected output of the
synchronous is fed through a Moseley Model 60D logarithmic
converter into an X-Y plotter which plots the measured insertion
loss in db as a function of frequency.

3. CONCLUSIONS

Dissipative transmission line filters have been shown to be an effective means of producing a low-pass transmission function with a uniformly high stop-band attenuation. The combination of these dissipative filters and conventional reactive filters can provide low-pass and bandpass filters with superior stop-band performance. Using the materials and procedures detailed in this report, practical filters can be constructed at a sufficiently low cost to warrant commercial application.

The dielectric losses of several iron and ferrite materials are sufficiently low at HF so that practical resonators may be constructed for use in the HF range. The ferrite filled helical resonator offers a considerable size reduction over a conventional resonator and operates satisfactorily at low power levels. At power levels above 0 dbm, the change in the characteristics of the ferrite material with power level makes it unsuitable for use as the dielectric material in a resonator. For power levels above 0 dbm, the carbonyl iron material offers a substantial improvement over ferrite in maintaining its dielectric properties independent of RF power level.

The results of sheilding effectiveness measurements of the iron and ferrite dielectric material indicate that these materials do not exhibit any exceptional shielding properties.

A rapid insertion loss measuring technique has been developed which permits direct plotting of insertion loss versus frequency for attenuations exceeding 60 db.

4. BIBLIOGRAPHY

1. Radar Design Corporation Catalog, 1963, p. 3, 105 Pickard Drive, Syracuse, New York.

2. L. Young, et al, "High Power Microwave Filters for the Suppression of Spurious Energy," Stanford Research Institute, pp. 49-50, February 1965, RADC-TR-64-515.

3. J. Fischer and J. Senn, "A New Family of Absorptive-Reactive RFI Filters," Proc. of Eighth Tri-Service Conference on Electromagnetic Compatibility, pp. 590-622, October 1962.

4. J. F. Clines, "UHF and VHF Narrow-Band Filter Characteristics," Stanford Research Institute Electron's Laboratory Technical Note No. 5, Contract AF 30(602)-2392, February 1962.

APPENDIX A

COMPLEX PERMEABILITY AND PERMITTIVITY

The electric flux density, $\bar{D}$, inside a dielectric material is related to the dielectric constant or permittivity, ϵ, of the material by

$$\bar{D} = \epsilon \bar{E} \quad . \tag{1}$$

For $\epsilon > \epsilon_o$, $\bar{D}$ can be written as

$$\bar{D} = \epsilon_o \bar{E} + \bar{P} \quad . \tag{2}$$

The additional electric flux density is produced by the polarization, $\bar{P}$, of the dielectric material. The polarization can be visualized as being created by the alignment of N elementary electric dipoles each with a dipole moment, $\bar{\mu}$, and producing a polarization of

$$\bar{P} = N \bar{\mu} \quad . \tag{3}$$

For a linear dielectric, the dipole moment is proportional to the applied $\bar{E}$ field. This proportionality can be expressed as

$$\mu = \alpha_p \bar{E} \quad , \tag{4}$$

substituting (4) in (3) gives

$$\overline{P} = N \, \alpha_p \, \overline{E} \quad . \tag{5}$$

For a time varying electric field, the polarization must also be time varying. However, for sufficiently rapid variation of the applied electric field, the alignment of the dipole moments may not follow the field instantaneously, and there will be a time lag between the internal polarization and the applied electric field. Consequently, for a time variation of

$$\overline{E} = \overline{E}_o \, e^{j\omega t} \quad , \tag{6}$$

the effect of the time lag can be taken into account by writing the polarization as

$$\overline{P} = N \, \alpha_p \, \overline{E}_o \, e^{j(\omega t - \theta)} \quad . \tag{7}$$

The resulting electric flux density is then given by substituting (7) in (2) as

$$D = \epsilon_o \, \overline{E}_o \, e^{j\omega t} + N \, \alpha_p \, \overline{E}_o \, e^{j(\omega t - \theta)} \quad , \tag{8}$$

which simplifies to

$$D = (\epsilon_o + N \alpha_p e^{-j\theta}) \; E_o \; e^{j\omega t} \quad , \tag{9}$$

or

$$D = (\epsilon_o + N \alpha_p e^{-j\theta}) \; \overline{E} \quad . \tag{10}$$

The quantity, $(\epsilon_o + N \alpha_p e^{-j\theta})$ is a complex number and, by analogy with Equation (1), can be considered as the complex permittivity of the dielectric material, i.e.

$$\overline{D} = \epsilon^* \; \overline{E} \quad . \tag{11}$$

A similar procedure can be used to illustrate the complex permeability of the dielectric material. The magnetic flux density, $\overline{B}$, is related to the magnetic field, $\overline{H}$, by

$$\overline{B} = \mu_o \; \overline{H} + \mu_o \; \overline{M} \quad . \tag{12}$$

The magnetization, $\overline{M}$, of the dielectric is the magnetic dipole moment per unit volume and, for linear dielectrics, is proportional to the applied magnetic field. Consequently,

$$\overline{M} = N \overline{m} = N \alpha_m \overline{H} \quad , \tag{13}$$

where N is the number of dipoles per unit volume, $\overline{m}$ is the magnetic dipole moment of an elementary dipole, and α_m is the proportionality constant.

Substituting (13) in (12) allows the magnetic flux density to be written as

$$B = \mu_o \, H_o \, e^{j\omega t} + N \, \alpha_m \, H_o \, e^{j(\omega t - \theta)} \quad , \tag{14}$$

or

$$\bar{B} = \mu_o \, \bar{H} \, (1 + \alpha_m \, Ne^{-j\theta}) \quad . \tag{15}$$

The term, $\mu_o \, (1 + \alpha_m \, Ne^{-j\theta})$, is the complex permeability, μ^*, and the flux density is

$$\bar{B} = \mu^* \, \bar{H} \quad . \tag{16}$$

The wave equations of the electromagnetic field in the dielectric material may be written in terms of these complex permeability and permittivity parameters as

$$\nabla^2 \, \bar{E} = \epsilon^* \, \mu^* \, \frac{\partial^2 \bar{E}}{\partial t^2} \quad , \tag{17}$$

and

$$\nabla^2 \, \bar{H} = \epsilon^* \, \mu^* \, \frac{\partial^2 \bar{H}}{\partial t^2} \quad . \tag{18}$$

If it is assumed that

$$\overline{H} = H(x) e^{j\omega t} \quad , \tag{19}$$

and

$$\overline{E} = E(x) e^{j\omega t} \quad , \tag{20}$$

then solutions for $\overline{E}$ and $\overline{H}$ describing a TEM wave propagating in the plus x direction are

$$\overline{E} = E_o e^{j\omega t - \gamma x} \quad , \tag{21}$$

and

$$\overline{H} = H_o e^{j\omega t - \gamma x} \quad , \tag{22}$$

where γ is the complex propagation factor

$$\gamma = j\omega \sqrt{\epsilon^* \mu^*} = \alpha + j\beta \quad , \tag{23}$$

and α and β are the attenuation and phase constants respectively.

The complex permeability and permittivity can be written as

$$\mu^* = \mu' - j\mu'' = |\mu| e^{j\theta_\mu} \quad , \tag{24}$$

125

and

$$\epsilon^* = \epsilon' - j\epsilon'' = |\epsilon| e^{j\theta_\epsilon} \quad , \tag{25}$$

where

$$\theta_\mu = \tan^{-1} \left(\frac{\mu''}{\mu'} \right) \quad , \tag{26}$$

and

$$\theta_\epsilon = \tan^{-1} \left(\frac{\epsilon''}{\epsilon'} \right) \quad . \tag{27}$$

Substituting in (23) gives

$$\gamma = j\omega \sqrt{|\mu| |\epsilon| e^{j(\theta_\mu + \theta_\epsilon)}} \quad , \tag{28}$$

Separating (28) into its real and imaginary parts gives

$$\gamma = j\omega \sqrt{|\mu| |\epsilon|} \left\{ \cos \frac{(\theta_\mu + \theta_\epsilon)}{2} + j \sin \frac{(\theta_\mu + \theta_\epsilon)}{2} \right\} \quad , \tag{29}$$

or

$$\gamma = \omega \sqrt{|\mu| |\epsilon|} \left\{ \sin \frac{(\theta_\mu + \theta_\epsilon)}{2} - j \cos \frac{(\theta_\mu + \theta_\epsilon)}{2} \right\} \quad . \tag{30}$$

The desired attenuation constant, α, is the real part of γ so that

$$\alpha = \omega \sqrt{|\mu| |\epsilon|} \sin \left(\frac{\theta_\mu + \theta_\epsilon}{2} \right) \quad . \tag{31}$$

The published data on dielectric materials is usually given in terms of the tangents of the angles θ_μ and θ_ϵ, and in terms of μ' and ϵ' rather than $|\mu|$ and $|\epsilon|$. Consequently, a more useful form of (31) is one which expresses the attenuation constant, α, in terms of these normally specified quantities. Let

$$\delta_\mu = \tan \theta_\mu \quad , \tag{32}$$

and

$$\delta_\epsilon = \tan \theta_\epsilon \quad . \tag{33}$$

Also

$$\sin \left(\frac{\theta_\mu + \theta_\epsilon}{2} \right) = \left[\frac{1 - \cos (\theta_\mu + \theta_\epsilon)}{2} \right]^{1/2} \quad , \tag{34}$$

which can be further expanded to give

$$\sin \left(\frac{\theta_\mu + \theta_\epsilon}{2} \right) = \left[\frac{1 - \cos \theta_\mu \cos \theta_\epsilon + \sin \theta_\mu \sin \theta_\epsilon}{2} \right]^{1/2} . \tag{35}$$

However, from (32) and (33),

$$\sin \theta_\mu = \frac{\delta_\mu}{\sqrt{1 + \delta_\mu^2}}$$

$$\sin \theta_\epsilon = \frac{\delta_\epsilon}{\sqrt{1 + \delta_\epsilon^2}}$$

$$\cos\,\theta_\mu = \frac{1}{\sqrt{1+\delta_\mu^2}}$$

$$\cos\,\theta_\epsilon = \frac{1}{\sqrt{1+\delta_\epsilon^2}} \qquad . \tag{36}$$

Substituting (36) in (35) yields

$$\sin\left(\frac{\theta_\mu + \theta_\epsilon}{2}\right) = \frac{1}{\sqrt{2}}\left[1 - \frac{1}{\sqrt{1+\delta_\mu^2}\,\sqrt{1+\delta_\epsilon^2}} + \frac{\delta_\mu\,\delta_\epsilon}{\sqrt{1+\delta_\mu^2}\,\sqrt{1+\delta_\epsilon^2}}\right]^{1/2}$$

$$= \frac{1}{\sqrt{2}}\left[\frac{\sqrt{1+\delta_\mu^2}\,\sqrt{1+\delta_\epsilon^2}\, - 1 + \delta_\mu\,\delta_\epsilon}{\sqrt{\left(1+\delta_\mu^2\right)\left(1+\delta_\epsilon^2\right)}}\right]^{1/2} \qquad . \tag{37}$$

Now

$$|\mu| = \frac{\mu'}{\cos\theta_\mu} = \mu'\sqrt{1+\delta_\mu^2} \qquad , \tag{38}$$

and

$$|\epsilon| = \frac{\epsilon'}{\cos\theta_\epsilon} = \epsilon'\sqrt{1+\delta_\epsilon^2} \qquad . \tag{39}$$

Substituting these values in the expression for $\sqrt{|\mu||\epsilon|}$ gives

128

$$\sqrt{|\mu||\epsilon|} = \left(\sqrt{\mu'\epsilon'}\right)\left[\sqrt{\left(1+\delta_\mu^2\right)\left(1+\delta_\epsilon^2\right)}\right]^{1/2} . \tag{40}$$

Substituting (37) and (40) in (31) gives for α,

$$\alpha = \omega \frac{\left(\sqrt{\mu'\epsilon'}\right)}{\sqrt{2}} \left[\sqrt{\left(1+\delta_\mu^2\right)\left(1+\delta_\epsilon^2\right)} - 1 + \delta_\mu \delta_\epsilon\right]^{1/2} . \tag{41}$$

Making the additional substitutions

$$\omega = 2\pi f$$

$$\mu' = \mu_0 \mu'_r$$

$$\epsilon' = \epsilon_0 \epsilon'_r , \tag{42}$$

gives an expression for α of

$$\alpha = (1.48)(10^{-8}) f \sqrt{\mu'_r \epsilon'_r}\left[\sqrt{\left(1+\delta_\mu^2\right)\left(1+\delta_\epsilon^2\right)} - 1 + \delta_\mu \delta_\epsilon\right]^{1/2} \text{ nepers m,} \tag{43}$$

which reduces to the desired relation of

$$\alpha = (12.85)(10^{-10}) f \sqrt{\mu'_r \epsilon'_r}\left[\sqrt{\left(1+\delta_\mu^2\right)\left(1+\delta_\epsilon^2\right)} - 1 + \delta_\mu \delta_\epsilon\right]^{1/2} \text{ db/cm.} \tag{44}$$

APPENDIX B

DERIVATION OF REFLECTION LOSS CORRECTION

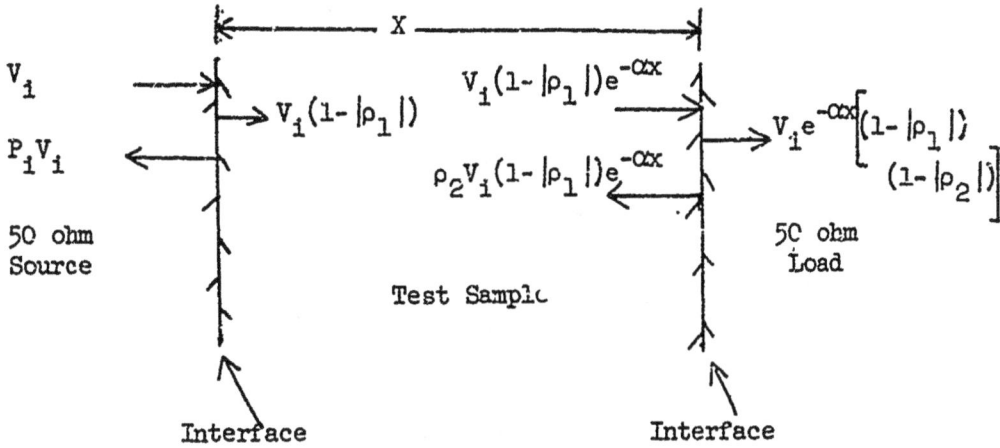

Figure 1. Measurement System

For the purpose of the derivation, the insertion loss measuring system with the test sample in place is represented as shown in Figure 1. The incident voltage wave, V_i, is partially reflected at the first interface and the component transmitted into the test sample in terms of the incident voltage and the reflection coefficient, ρ_1, is

$$V_{trans} = V_i(1 - |\rho_1|) \quad . \tag{1}$$

In traversing the test sample the wave reaching the output interface is attenuated by the factor, $e^{-\alpha x}$, where α is the attenuation constant of the material and x is the length of the test sample. A second partial reflection takes place at the output interface to produce a net output of

$$V_{out} = V_1 e^{-\alpha x}(1 - |\rho_1|)(1 - |\rho_2|) \quad .$$

(2)

Equation (2) ignores the contributions to the output produced by multiple internal reflections. These multiple reflections are attenuated by the factor, $e^{-\alpha x}$, each time they traverse the test sample as well as by the fact that only a fraction of each incident wave is reflected through the test sample from each encounter with an interface. For the materials tested, the values of ρ and α were such that these multiple reflections do not contribute significantly to the output voltage wave. The total insertion loss of the sample is then given by the ratio

$$L_{total} = \frac{V_1}{V_{out}} = \frac{1}{e^{-\alpha x}(1 - |\rho_1|)(1 - |\rho_2|)} \quad ,$$

(3)

in which the factor, $e^{\alpha x}$, is the dissipation loss of the test material. The reflection loss factor is

$$L_R = \frac{1}{(1 - |\rho_1|)(1 - |\rho_2|)}$$

(4)

131

In the 50 ohm test system used, the reflection coefficients are given in terms of the characteristic impedance, Z_t, of the test sample by

$$\rho_1 = \frac{Z_t - 50}{Z_t + 50} \quad , \tag{5}$$

and

$$\rho_2 = \frac{50 - Z_t}{50 + Z_t} \quad . \tag{6}$$

Therefore, from Equation (4)

$$L_R = \frac{1}{\left(1 - \left|\frac{Z_t - 50}{Z_t + 50}\right|\right)\left(1 - \left|\frac{50 - Z_t}{50 + Z_t}\right|\right)} \quad , \tag{7}$$

which simplifies to

$$L_R = \frac{(Z_t + 50)^2}{(100)(2\ Z_t)} \quad . \tag{8}$$

The characteristic impedance of the test specimen is a function of the dimensions of the coaxial line and the μ_r^* and ϵ_r^* of the test material and can be expressed as

$$Z_t = \left(\sqrt{\mu_r^*/\epsilon_r^*}\right)\ 60\ \ln\left(\frac{d_1}{d_2}\right) \quad , \tag{9}$$

where d_1 and d_2 are the diameters of the inner and outer conductors

132

of the coaxial line. Substituting the values, $d_1 = 0.102"$ and $d_2 = 0.561"$, gives

$$Z_t = 100 \sqrt{\mu_r^* / \epsilon_r^*} \quad .$$
(10)

Substituting (10) in (8) and simplifying gives

$$L_R = \frac{\left(\sqrt{\mu_r^* / \epsilon_r^*} + \frac{1}{2} \right)^2}{2 \sqrt{\mu_r^* \epsilon_r^*}} \quad .$$
(11)

Expressing L_R in decibels gives

$$L_R = 40 \log_{10} \left(\sqrt{\mu_r^* / \epsilon_r^*} + \frac{1}{2} \right) - 10 \log_{10} \left(4 \mu_r^* / \epsilon_r^* \right) \quad .$$
(12)

APPENDIX C

TAPER DESIGN PROCEDURE

The objective of the taper design procedure is to determine the necessary dimensions of the center conductor of a coaxial line to provide a fixed characteristic impedance at every point along the taper. It is assumed that the length of the taper is given. The taper length used is not critical and need only be of sufficient length to dissipate the internal reflections at frequencies where the μ_r^* and ϵ_r^* of the dielectric material differ from the values at the design frequency.

In developing the necessary taper design equations, the initial problem is that of calculating the impedance of a coaxial line partially filled with a dielectric material.

Referring to Figure 1, Radius r_3 is known, being the chosen dimension of the coaxial line. Radius r_2 is the distance from the center of the line to the air-dielectric boundary. Radius r_1 is the dimension of the center conductor which is to be determined.

Assuming TEM mode of propagation inside the lossy dielectric, the per unit inductance and capacitance are

$$L = \frac{\mu_0}{2\pi} \left[\mu_r \ln \frac{r_2}{r_1} + \ln \frac{r_3}{r_2} \right] \quad , \tag{1}$$

and

$$C = \frac{2\pi \epsilon_0 \epsilon_r}{\ln \frac{r_2}{r_1} + \epsilon_r \ln \frac{r_3}{r_2}} \quad , \tag{2}$$

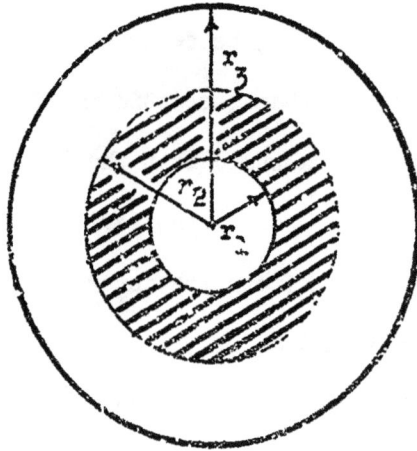

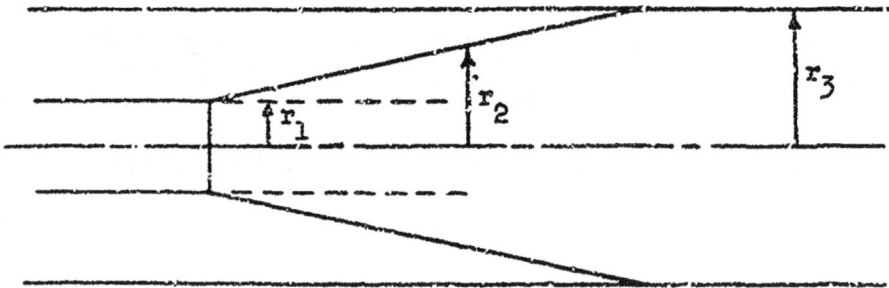

Figure 10. Taper Outline

135

where ϵ_r and μ_r are the relative permittivity and permeability respectively.

The characteristic impedance of the line is

$$Z_o = \sqrt{L/C} \quad . \tag{3}$$

Substituting (1) and (2) in (3) gives

$$Z_o = \left[\frac{\mu_o}{2\pi}\left(\mu_r \ln \frac{r_2}{r_1} + \ln \frac{r_3}{r_2}\right)\left(\ln \frac{r_2}{r_1} + \epsilon_r \ln \frac{r_3}{r_2}\right)\left(\frac{1}{2\pi \epsilon_o \epsilon_r}\right)\right]^{1/2}, \tag{4}$$

but

$$\frac{\mu_o}{4\pi\epsilon_o} = 60 \quad . \tag{5}$$

Therefore,

$$Z_o = 60\left[\frac{1}{\epsilon_r}\left(\mu_r \ln \frac{r_2}{r_1} + \ln \frac{r_3}{r_2}\right)\left(\ln \frac{r_2}{r_1} + \epsilon_r \ln \frac{r_3}{r_2}\right)\right]^{1/2}. \tag{6}$$

Further,

$$\epsilon_r \left(\frac{Z_o}{60}\right)^2 = \left(\mu_r \ln \frac{r_2}{r_1} + \ln \frac{r_3}{r_2}\right)\left(\ln \frac{r_2}{r_1} + \epsilon_r \ln \frac{r_3}{r_2}\right) , \tag{7}$$

or

$$\epsilon_r \left(\frac{Z_o}{60}\right)^2 = \mu_r \left(\ln \frac{r_2}{r_1}\right)\left(\ln \frac{r_2}{r_1}\right) + \left(\ln \frac{r_3}{r_2}\right)\left(\ln \frac{r_2}{r_1}\right)$$

$$+ \left(\epsilon_r \mu_r\right)\left(\ln \frac{r_3}{r_2}\right)\left(\ln \frac{r_2}{r_1}\right) + \epsilon_r \left(\ln \frac{r_3}{r_2}\right)\left(\ln \frac{r_3}{r_2}\right) \quad . \tag{8}$$

Let

$$X = \ln \frac{r_2}{r_1} \text{ and } K = \ln \frac{r_3}{r_2} \quad . \tag{9}$$

Making these substitutions in (6) gives

$$\epsilon_r \left(\frac{Z_o}{60}\right)^2 = \mu_r X^2 + \left(1 + \epsilon_r \mu_r\right) KX + \epsilon_r K^2 \quad . \tag{10}$$

Rearranging into the usual quadratic form,

$$X^2 + \left[\frac{(1 + \epsilon_r \mu_r) K}{\mu_r}\right] X + \left[\frac{\epsilon_r}{\mu_r}\right] \left[K^2 - \left(\frac{Z_o}{60}\right)^2\right] = 0 \quad , \tag{11}$$

and solving for X gives

$$X = \frac{-\left(1 + \epsilon_r \mu_r\right)}{2\mu_r} K + \frac{1}{2} \left[\left(\frac{1 + \epsilon_r \mu_r}{\mu_r}\right)^2 K^2 - \frac{4\epsilon_r}{\mu_r} K^2 + \frac{\epsilon_r}{900 \, \mu_r} Z_o^2\right]^{1/2} . \tag{12}$$

The values of X are obtained by substituting appropriate values into the right hand side of the equation. It is assumed that the values of μ_r and ϵ_r are either given or have been measured.

The value of K is computed by first measuring r_2 which may be done graphically or algebraically (for simple linear tapers). With r_2 and r_3 available, K follows directly. Note that r_2 varies along the length of the taper. Consequently, K varies which determines the different values of X along the taper.

137

Z_o is the characteristic impedance desired and is chosen to match the remainder of the transmission system.

Once X has been computed for a particular point along the taper, the corresponding center conductor radius, r_1, immediately follows from

$$r_1 = \frac{r_2}{\ln^{-1} X} ,$$ (13)

APPENDIX D

HIGH POWER AMPLIFIER

To facilitate the testing of HF filters, a power amplifier was designed and constructed which was capable of delivering 100 watts at HF into a 50 ohm load. The amplifier consisted of a 6CL6 driver stage and a 4X150A Class C power output amplifier as shown in the schematic of Figure 1 A 50 ohm source capable of supplying three volts rms, such as the HP 606A, is used to drive the power amplifier. The amplifier operates over the frequency range of 6.5 to 11 Mc on the low band and 10 to 18 Mc on the high band.

The input RF signal drives the grid circuit of the 6CL6 through a tap on the tuned grid circuit. The 6CL6 driver operates Class A, and its amplified output is developed across the tuned circuit formed by L1 and C2. This resonant circuit serves as the plate load for the 6CL6 and the tuned grid circuit of the 4X150A. Considerable care in shielding the grid circuit of the 6CL6 as well as careful grounding of the screen circuit were necessary to prevent oscillation in the driver stage.

The control grid of the 4X150A is biased at -150 volts which is approximately 50 volts below cutoff. The RF voltage developed across the resonant circuit of L1 and C2 is sufficient to drive the 4X150A into conduction. A Pi network is used to match the plate impedance of the 4X150A grid to a 50 ohm load. The meter

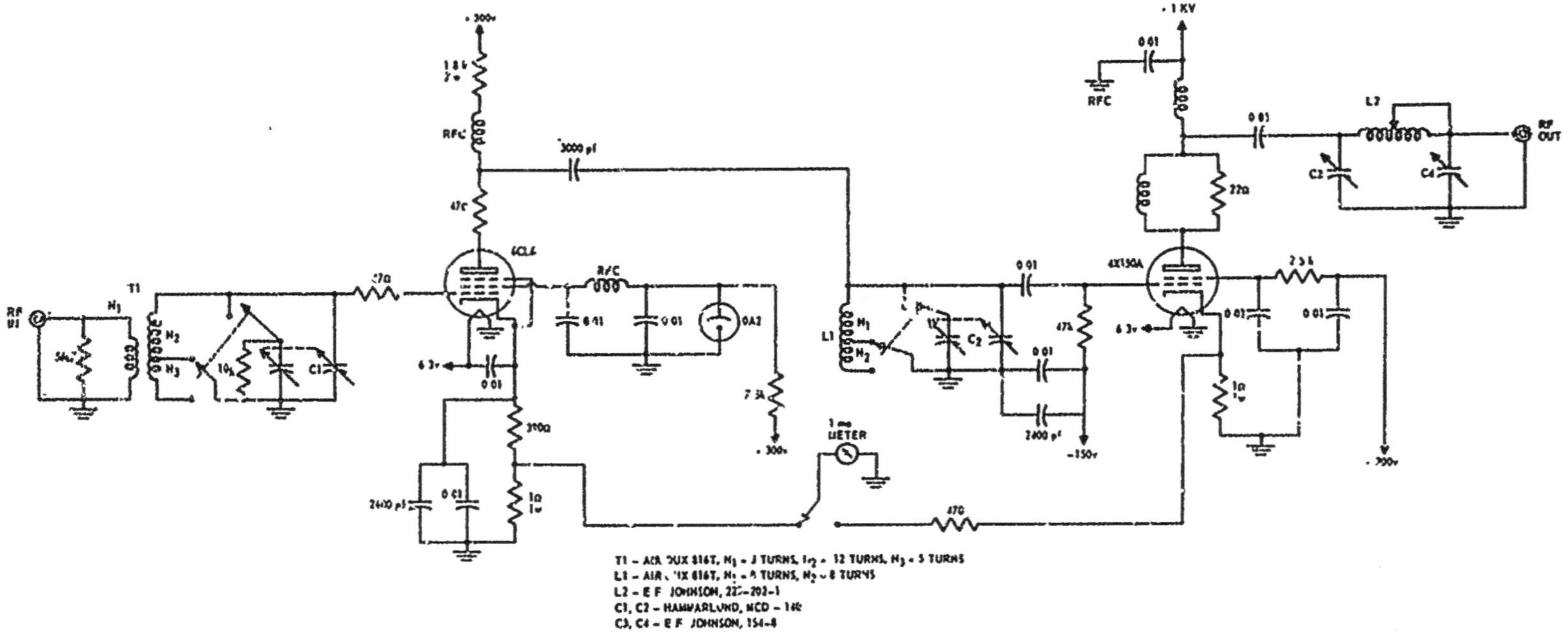

Figure 1D. High Power Amplifier.

circuit, which monitors the cathode current of each tube, is used as an aid in tuning the amplifier.

www.ingramcontent.com/pod-product-compliance
Lightning Source LLC
Chambersburg PA
CBHW080555220326
41599CB00032B/6486